Quantum Computing for Computer Architects

Second Edition

Synthesis Lectures on Computer Architecture

Editor

Mark D. Hill, *University of Wisconsin*

Synthesis Lectures on Computer Architecture publishes 50- to 100-page publications on topics pertaining to the science and art of designing, analyzing, selecting and interconnecting hardware components to create computers that meet functional, performance and cost goals. The scope will largely follow the purview of premier computer architecture conferences, such as ISCA, HPCA, MICRO, and ASPLOS.

Quantum Computing for Computer Architects, Second Edition
Tzvetan S. Metodi, Arvin I. Faruque, and Frederic T. Chong
2011

Processor Microarchitecture: An Implementation Perspective
Antonio González, Fernando Latorre, and Grigorios Magklis
2010

Transactional Memory, 2nd edition
Tim Harris, James Larus, and Ravi Rajwar
2010

Computer Architecture Performance Evaluation Methods
Lieven Eeckhout
2010

Introduction to Reconfigurable Supercomputing
Marco Lanzagorta, Stephen Bique, and Robert Rosenberg
2009

On-Chip Networks
Natalie Enright Jerger and Li-Shiuan Peh
2009

The Memory System: You Can't Avoid It, You Can't Ignore It, You Can't Fake It
Bruce Jacob
2009

Quantum Computing for Computer Architects, Second Edition

Tzvetan S. Metodi, Arvin I. Faruque, and Frederic T. Chong

ISBN: 978-3-031-00603-6   paperback
ISBN: 978-3-031-01731-5   ebook

DOI 10.1007/978-3-031-01731-5

A Publication in the Springer series
SYNTHESIS LECTURES ON ADVANCES IN AUTOMOTIVE TECHNOLOGY

Lecture #13
Series Editor: Mark D. Hill, *University of Wisconsin*
Series ISSN
Synthesis Lectures on Computer Architecture
Print 1935-3235   Electronic 1935-3243

Quantum Computing for Computer Architects

Second Edition

Tzvetan S. Metodi
The Aerospace Corporation

Arvin I. Faruque
University of California, Santa Barbara

Frederic T. Chong
University of California, Santa Barbara

SYNTHESIS LECTURES ON COMPUTER ARCHITECTURE #13

ABSTRACT

Quantum computers can (in theory) solve certain problems far faster than a classical computer running any known classical algorithm. While existing technologies for building quantum computers are in their infancy, it is not too early to consider their scalability and reliability in the context of the design of large-scale quantum computers. To architect such systems, one must understand what it takes to design and model a balanced, fault-tolerant quantum computer architecture. The goal of this lecture is to provide architectural abstractions for the design of a quantum computer and to explore the systems-level challenges in achieving scalable, fault-tolerant quantum computation.

In this lecture, we provide an engineering-oriented introduction to quantum computation with an overview of the theory behind key quantum algorithms. Next, we look at architectural case studies based upon experimental data and future projections for quantum computation implemented using trapped ions. While we focus here on architectures targeted for realization using trapped ions, the techniques for quantum computer architecture design, quantum fault-tolerance, and compilation described in this lecture are applicable to many other physical technologies that may be viable candidates for building a large-scale quantum computing system. We also discuss general issues involved with programming a quantum computer as well as a discussion of work on quantum architectures based on quantum teleportation. Finally, we consider some of the open issues remaining in the design of quantum computers.

KEYWORDS

quantum computing, computer architecture, fault tolerance, error correction, trapped ions, teleportation, qubit, quantum logic array, quantum simulation, quantum algorithms

Contents

Preface

Quantum computers can (in theory) solve certain problems far faster than a classical computer running any known classical algorithm. While existing technologies for building quantum computers are in their infancy, it is not too early to consider their scalability and reliability in the context of the design of large-scale quantum computers. To architect such systems, one must understand what it takes to design and model a balanced, fault-tolerant quantum computer architecture. The goal of this lecture is to provide architectural abstractions for the design of a quantum computer and to explore the systems-level challenges in achieving scalable, fault-tolerant quantum computation.

In this lecture, we provide an engineering-oriented introduction to quantum computation with an overview of the theory behind key quantum algorithms. Next, we look at architectural case studies based upon experimental data and future projections for quantum computation implemented using trapped ions. While we focus here on architectures targeted for realization using trapped ions, the techniques for quantum computer architecture design, quantum fault-tolerance, and compilation described in this lecture are applicable to many other physical technologies that may be viable candidates for building a large-scale quantum computing system. We also discuss general issues involved with programming a quantum computer as well as a discussion of work on quantum architectures based on quantum teleportation. Finally, we consider some of the open issues remaining in the design of quantum computers.

The second edition contains new material intended to both provide the reader with a deeper background in quantum computation as well as novel concepts and results in quantum computer architecture. A new chapter (Chapter 3) augments the introduction to quantum computation in Chapter 2 with detailed expositions of key quantum algorithms, including Shor's algorithm for factoring integers in polynomial time and Grover's algorithm for quantum search. Chapter 7 has been expanded with recent work on ion-trap quantum computer architectures. Additionally, we have added another new chapter (Chapter 8) which provides a case study of the an application of the architecture described in Chapter 7 to a quantum simulation task.

The book begins with a brief background in Chapter 2 which compares the basic operations for quantum computation to the conventional computing scheme by focusing on computation rather than physics. We describe, in some detail, the concept of qubits, quantum logic gates, and other important components for quantum computing relevant to the circuit model for quantum computation. Following this, we present an overview the of mathematical principles behind of a few notable quantum algorithms in Chapter 3. In Chapter 4, we introduce three high-level requirements for a scalable quantum architecture and describe each requirement independently in the following sections: reliable implementation technology in Section 4.1, efficient error correction schemes in Section 4.2, and efficient quantum resource distribution in Section 4.3. Modeling and simulating quantum computational structures and cycle-level quantum simulation methods are described in Chapter 5, including a brief introduction of the stabilizer formalism for quantum computation and error correction. A set of architectural elements for a quantum architecture is described in Chapter 6. The concept of *quantum memory hierarchy* is described in Section 6.2. In Chapter 7, we give a case study for a quantum architecture, the Quantum Logic Array (QLA), based on our previous work [60, 140]. Chapter 8 provides a treatment of how we can program a quantum architecture, and Chapter 9 covers the QLA architecture for a quantum simulation application. Chapter 10 offers a discussion into the alternate methods for achieving fault-tolerant universal quantum logic, namely, performing quantum operation through the concept of teleportation. Finally, we conclude with Chapter 11, where we give a brief summary of what we have done.

The authors would like to thank our research collaborators, Darshan Thaker, Jedidiah Crandall, John Oliver, Andrew Cross and Professor Isaac Chuang for discussing much of the material with us and, most importantly, providing us with useful criticism for the content of this book. We would like to thank MIT ion trappers Ken Brown, Jaroslaw Labaziewicz, and Rob Clark for answering technical questions about the physical implementations of quantum computers. Rodney Van Meter comments were invaluable in the later stages of writing the first edition. Discussions at UCSB about quantum architecture and algorithms with Chris Bunch, Brian Drawert, and Wim Van Dam helped form a portion of the new material in the second edition. We would also like to thank Andrew Cross and Margaret Martonosi for their comments on the second edition of this book. And finally, we thank Mark Hill from the University of Wisconsin Madison for making this book possible, providing us with the necessary guidance for the its content, and also supplying us with additional comments on the second edition.

Tzvetan S. Metodi, Arvin I. Faruque, and Frederic T. Chong
March 2011

CHAPTER 1

Introduction

Quantum computation may seem to be a topic for science fiction, but small quantum computers have existed for several years [212] and larger machines are on the drawing table [114]. These efforts have been fueled by a tantalizing property: while conventional computers employ a binary representation that allows the amount of processed information to scale linearly with resources, at best, quantum computations employ quantum phenomena that that can interact to allow the amount processed information to scale exponentialy in the number of 'quantum bits' in the system. Architecting large scale systems to exploit this potential is the focus of this book. Our goal is to provide architectural abstractions common to potential technologies and explore the systems-level challenges in achieving scalable, fault-tolerant quantum computation. While quantum technologies are in their infancy, we shall see that it is not too early to consider scalability and reliability. In fact, such considerations are critical to guide the development of viable device technologies.

The premise of this book is directed at quantum computation (QC) architectural issues. We underscore the fact that the basic tenet of *large-scale* quantum computing is *reliability through system balance* - the need to protect and control the quantum information just long enough for the algorithm to complete execution. To architect QC systems, one must understand what it takes to design and model a balanced, fault-tolerant quantum architecture, just as the concept of balance drives conventional architectural design. For example, the number of functional units in classical processors is matched to the register file depth or the memory bandwidth to the cache miss rate.

We illustrate the application of the concept of balance to quantum architecture design though a comparison of two quantum computer architectures designed to implement Shor's quantum factoring algorithm: the Quantum Logic Array (QLA) architecture [140] and its successor, the Compressed Quantum Logic Array (CQLA) architecture [60]. Both architectures are homogeneous, tiled architectures designed for the circuit-based model of computation. Both architectures can be described as a grid of computational tiles, where each tile represents a single fault-tolerant quantum bit, or a *logical qubit*. A logical qubit is encoded into a number of physical qubits such that its underlying geometrical construction is intended to provide the necessary resources for quantum error correction (by far the most dominant operation for a quantum computer [153]). The primary difference between the two architectures is that, in the CQLA, the authors have created a more balanced system-level organization of the underlying qubit resources by trading some of the excess reliability for gains in the chip area and the speed of the architecture.

This trade in reliability for gain in speed and area is illustrated using the performance design pyramid in Figure 1.1. The pyramid has three high-level performance axes: Area, Speed, and Reliability. Improvement for each of the three performance parameters is in the direction of the pyramid apex. There is also a minimum reliability mark for each logical operation within the quantum computer, under which the factoring of integers larger than 2048 bits is not possible [60, 140]. There are, perhaps, similar constraints along the other two axes of the pyramid, which depend on the set of applications or on technology limitations. Such constraints have had some exploration in the literature [103] [220].

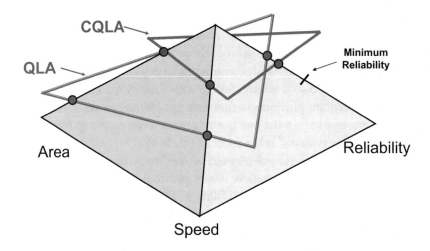

Figure 1.1: The quantum architecture performance design pyramid, illustrating the relative performance design space for the QLA and CQLA quantum architectures. Improvement is in the direction of the pyramid apex, and there is a minimum reliability mark for each operation, below which computationally relevant quantum applications will fail.

We hope that our description of specific architecture models will enable the reader to continue the advancement of scalable quantum architecture research and to tackle some of the key open questions. For example, what is the best way to integrate fault-tolerant scalable data storage structures, computational structures, scalable communication mechanisms, and classical schedulers that orchestrate the program execution? When finished, the reader should be able to identify the different tradeoffs between the various requirements for scalable quantum computation, and most importantly (via clever system design), be able to create a quantum architecture that balances reliability, area, and time performance such that it is relevant and within the reach of future technological advances.

Unfortunately, building large-scale quantum machines is extremely difficult. This difficulty is consistent with the intuition that quantum information in nature can be controlled only in very small, carefully isolated physical systems such as single photons and atoms. Binary information in quantum mechanical systems can be stored in a single fragile unit of quantum data (known as a *qubit* [177]) in the distinct energy states of an atom, or the polarization states of a photon. On the other hand, larger-scale systems naturally couple with the environment and exhibit the behavior governed by classical physics that is so familiar to our everyday experiences. A corollary of this observation is that the physics of quantum computation often defies our classical intuition, and is responsible for both the potential power of QC and the difficulty in realizing reliable quantum computers.

Following Feynman's famous observation[1] about the significant gap between classical computational models and quantum mechanical ones, the first model in the context of a *quantum Turing machine* was introduced by Benioff [20]. Subsequently, Deutsch [63, 64] described the quantum circuit model as a universal simulator for the quantum Turing machine, with exponential overhead. Bernstein and Vazirani [26] followed Deutsch's work with the description of a universal quantum Turing machine constructed with only a polynomial overhead.

Since the construction of the universal quantum circuit model, the ability to control and manipulate quantum information through a sequence of gates has led to several quantum algorithms with substantial advantages over known algorithms using traditional computation. The most significant is Peter Shor's algorithm for factoring the product of two large primes in polynomial time [184]. [2] Additional quantum algorithms and applications include Grover's fast database search algorithm [89]; adiabatic solution of optimization problems [48]; precise clock synchronization [51]; quantum key distribution [24]; and recently, Gauss sums [211] and Pell's equation [91]. Commercially, quantum technologies have been shown to enable unconditionally secure communication,[3] leading to the creation of companies offering real products [81, 170].

A practical large-scale quantum computer that can utilize the full potential of these algorithms must be capable of reaching a system size of $S = KQ \geq 10^{12}$ logical operations, where K denotes the number of computational steps and Q denotes the number of computational units. The problem with sustaining such a large amount of quantum computation is that the quantum information carriers (the qubits) continuously interact with external noise sources in the environment and *decohere*, eventually losing their quantum data. In addition, different quantum states are not mutually exclusive, as different classical bitstrings are, and they may become entangled with one another. While entanglement is responsible for the power of quantum computation, it also enables errors to spread to the entire system of entangled qubits if care is not taken at the microarchitecture level to limit their spread.

A key theoretical breakthrough in *scalable* QC was the development of a theory of fault-tolerant quantum error correction (FTQEC). FTQEC allows the reliability of large systems to be arbitrarily increased by recursively encoding the state of a single logical qubit into the state of two or more lower-level qubits. To sustain reliable computation for an extended period of time using noisy physical gates, logic gates must be implemented directly over the logical qubits and be able to

---

[1]The observation was made in Feynman's talk during the First Conference on the Physics of Computation held at MIT in 1981. Feynman noted that it is impossible to simulate the evolution of a quantum system efficiently on a classical computer, unless an inherently quantum computer is constructed.

[2]The security of the widely used RSA public-key cryptosystem relies on the assumption that factoring large integers is very hard on conventional computers [172], where the best-known classical algorithms for factorization are super-polynomial [40].

[3]Mathematically, quantum key distribution (QKD) has been proven unbreakable [17, 23, 65, 134], but recent experimental observations have shown that the probabilistic nature of the protocols, coupled with noisy devices used to send and receive the quantum states, allow the attacker Eve, to break the system with high probability of success [147].

perform logic functions without decoding, in such a way that errors do not spread through the data interference patterns of the encoded states. Later in this lecture (Chapter 4.2.8), the implications of managing these challenging overheads versus the potential benefits of quantum computation will be discussed.

The physical implementation of large, redundant structures gives rise to another fundamental challenge: communication of quantum data across distances, which are non-trivial for current technologies. Consequently, one of the greatest challenges of designing a large, practically useful quantum computer is finding an architecture that incorporates the required amount of fault-tolerance, while minimizing communication and resources overhead. In the two example architectures, which we discuss in this book, the communication challenge is addressed by employing the following design principles: first, the idea that specializing architectural components into compute and memory regions is advantageous; second, the idea that the concept of quantum teleportation [21] is advantageous for communication across long distances in large-scale architectures; given these two design principles, we describe a general abstraction for scalable quantum architectures in which the dominant cost is communication between compute and memory regions, using teleportation. In such a system, a compiler infrastructure for the scheduling of quantum computations would maximize usage of data while it is in the compute region and minimize movement in and out to memory (analogous to minimizing register spilling in conventional processors).

The key ideas computer architects can take away after reading this book that are relevant for the system design of scalable quantum computers are:

- Achieving "good" system performance is synonymous with the realization of a workable balance between reliability, communication resources, and latency of computation in a quantum architecture.

- Quantum information cannot be cloned - thus, when transferred from source to destination along a quantum channel, it must be transferred in such a way that no trace is left at the source.

- Quantum information can be transported by physically moving the qubits, transferring the information to a shared medium such as a quantum bus or a secondary quantum system that allows efficient qubit movement, successive swapping between adjacent qubits, or through the concept of teleportation.

- The need for reliability allows for interesting match-ups between various system components. For example, communication and computation can be overlapped at the system level, due to the considerable resources and latency overhead spent on error correction during logic gate execution of encoded data.

- There are many different ways to implement universal quantum logic at the application level. Gates can be built into the communication protocols or applied to the quantum data in a traditional manner analogous to classical circuits.

- System balance and fault-tolerance depends on the balance between different encodings of the quantum data defined by the error correcting codes used across different regions in the architecture, where the cost and resources required to transfer from one region to another are carefully determined by the executed application.

- The memory hierarchy in classical computation is analogous to *code hierarchy* in quantum systems. The transfer from storage regions to computational regions may require transfer from one encoding of the data to another, not a transfer from one technology type to another.

CHAPTER 2

Basic Elements for Quantum Computation

This chapter presents a general overview of quantum computation that will provide the reader with an idea of how quantum data is stored and manipulated. We discuss how the notion of *quantum parallelism* – where a function $f(x)$ can be evaluated simultaneously for a number of inputs – is a somewhat oversimplified interpretation of the power of quantum computing since physical measurement of the system collapses the state vector to a single binary bitstring state. This means that even if all inputs to a function can be evaluated in one operation using a set of quantum qubit registers, the result for only one of those inputs can be observed, making quantum computing seem no more powerful than classcial computing. In this chapter, we describe how unique quantum information properties, such as quantum interference and quantum entanglement, preserve quantum parallelism for specific functions and for specific types of solutions to these functions that represent global properties, such as maxima, minima, or the function's period.

2.1 CLASSICAL VS. QUANTUM SIGNAL STATES (BITS VS. QUBITS)

Well-characterized signal states are one of the most fundamental requirements for realizing a digital computer. In classical computing, signal states are represented as binary bits, "0" or "1". During computation, the "1" bit is denoted as the presence of voltage through a silicon gate, while the "0" bit is marked as the lack of voltage. Thus, a classical bit exists in one of two well-defined states, "1" or "0".

On the other hand, the basic unit of state in quantum computers, the *qubit*, is described by quantum mechanical two-level systems, such as the two spin states of spin $1/2$ atoms, or the horizontal and vertical polarization states of a single photon. The difference between a qubit and a bit is that the physical state of a qubit is described by complex-valued amplitudes equal to the square root of finding the qubit in one of the two binary states "0" and "1". Similarly, the state of an n-qubit quantum system is desribed by 2^n complex-valued probability amplitudes, each equal to the probability of finding the quantum system into any of the 2^n possible n-bit binary bitstrings. Mathematically, the state of an n-qubit quantum system can be represented as a complex-valued 2^n-element vector. Furthermore, a single quantum gate (represented as a $2^n \times 2^n$ unitary matrix) applied to an n-qubit quantum system acts simultaneously on all 2^n elements of the system state vector. This means that the amount of information that can potentially be processed by quantum computers doubles with each addition qubit in the system.

The state of a single qubit, $|\Psi\rangle$, can be written as:

$$|\Psi\rangle \;=\; a|0\rangle + b|1\rangle \tag{2.1}$$

The amplitudes a and b are complex coefficients, such that $|a|^2 + |b|^2 = 1$, and the "$|\cdot\rangle$" notation, known as the *Dirac-Ket*, is used to denote a particular quantum state. The quantity $|a|^2$ is the probability that the qubit will be observed in the state $|0\rangle$, and similarly $|b|^2$ is the probability that the qubit exists in the state $|1\rangle$. Without direct observation, however, the state of a single qubit spans a two-dimensional vector space defined by the two-element complex valued vector $[a, b]^T$, where the most general single qubit state $|\Psi\rangle$ can be written in vector form as:

$$|\Psi\rangle \;=\; a|0\rangle + b|1\rangle \;=\; a\begin{bmatrix}1\\0\end{bmatrix} + b\begin{bmatrix}0\\1\end{bmatrix} \;=\; \begin{bmatrix}a\\0\end{bmatrix} + \begin{bmatrix}0\\b\end{bmatrix} \;=\; \begin{bmatrix}a\\b\end{bmatrix}$$

The state of a quantum computer with a storage total of two qubits is described by a four-dimensional vector space where each dimension can be distinguished by the four bit strings: $|00\rangle$, $|01\rangle$, $|10\rangle$, and $|11\rangle$. An arbitrary state of a two-qubit system (denoted as $|\Psi\rangle$), is described by a four element, complex valued vector $[c_0, c_1, c_2, c_3]^T$:

$$|\Psi\rangle \;=\; c_0|00\rangle + c_1|01\rangle + c_2|10\rangle + c_3|11\rangle, \tag{2.2}$$

and since the amplitudes are square roots of probabilities, the sum $|c_0|^2 + |c_1|^2 + |c_2|^2 + |c_3|^2$ must equal to unity. Similarly, three qubits will be in a superposition of eight bit strings, encoding the numbers zero through seven in each bit string. Thus, computing a function $f(x)$ where $x = 0, 1, ..., 7$ can potentially be done using three qubits, via a *single* logic gate.

In general, an n-qubit quantum system may represent 2^n bitstrings distinguished by 2^n complex valued coefficients:

$$|\Psi\rangle \;=\; \sum_{i=0}^{2^n-1} c_i|x_i\rangle, \quad such\ that \sum_{i=0}^{2^n-1} |c_i|^2 = 1, \tag{2.3}$$

where each x_i represents the i'th bitstring from 0 to 2^{n-1}. Because each additional qubit doubles the number of pure states (bitstrings) represented and subsequently manipulated by logic operations at each clock step, we can see how quantum computation has the potential to offer exponential scaling of the computing power with only a polynomial increase in the data resources.

2.2   LOGIC OPERATIONS AND CIRCUITS

A circuit in both classical and quantum computation is made up of *wires* and *logic gates*. The classical circuit model of computation is composed of boolean logic gates (such as the logically-universal

NAND gate) with a number of input and output bits which travel as an electric current, typically through copper wires. The actions of classical gates on bit strings are guided by the rules of boolean algebra and regardless of the gate type, the fundamental classical operation is a *bit-flip*: a decision whether the value of one or more bits in a register will be flipped.

A single gate in a quantum circuit with one or more input qubits in the initial state $|\Psi\rangle$ transforms the state to a different state $|\Psi'\rangle$ by changing all probability amplitudes that describe the state vector $[c_0, c_1, \cdots c_{n-1}]^T$ to $[c_0', c_1', \cdots c_{n-1}']^T$. Thus, quantum logic gates can be mathematically described as $2^n \times 2^n$ matrices. Since the square of the elements of the new state vector must also add up to unity (i.e., $\sum_{i=0}^{n-1} |c_i'|^2 = 1$), all quantum logic gates are unitary matrices which preserve the $p-$norm of a vector. An arbitrary operator U acts on the state $|\Psi\rangle$ as follows:

$$U|\Psi\rangle = U \begin{bmatrix} c_0 \\ c_1 \\ \vdots \\ c_{n-1} \end{bmatrix} = \begin{bmatrix} c_0' \\ c_1' \\ \vdots \\ c_{n-1}' \end{bmatrix} = |\Psi'\rangle, \tag{2.4}$$

Since the inverse of a unitary operator always exists, applying U^{-1} to $|\Psi'\rangle$ will restore the state back to $|\Psi\rangle$. This means that, unlike classical logic, quantum logic is *reversible* and every n-qubit input to a quantum operation must also have n output qubits.

The most general 2×2 operator U that acts on a single qubit is a rotation matrix, written as:

$$U = \begin{bmatrix} e^{i\alpha/2} & 0 \\ 0 & e^{-i\alpha/2} \end{bmatrix} \times \begin{bmatrix} \cos\theta/2 & \sin\theta/2 \\ -\sin\theta/2 & \cos\theta/2 \end{bmatrix} \times \begin{bmatrix} e^{i\beta/2} & 0 \\ 0 & e^{-i\beta/2} \end{bmatrix}, \tag{2.5}$$

where the values α, β, and θ denote the angles of rotation along the different degrees of freedom. A valid rotation of the state of a single qubit can be arbitrarily small; thus, there is an infinite number of possible operations that can be applied to a single qubit.

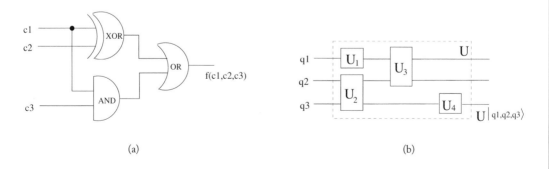

(a)   (b)

Figure 2.1: (a) An example classical circuit, where the output bit is a function of the input bits $\{c1, c2, c3\}$. (b) A 3-timestep quantum circuit. The notation U_i denotes an i-qubit operator.

The overall function of the sequence of operations in an entire n-qubit quantum circuit divided into K timesteps can be collectively described by a $2^n \times 2^n$ unitary operator U, where $U = U_k \times U_{k-1} \times ...U_1$. Each U_i is the $2^n \times 2^n$ unitary operator that describes the i'th timestep in the circuit and the collective action of the sequence is the product of all individual operators for each timestep. A schematic of a quantum and a classical circuit is shown in Figure 2.1. Figure 2.1(a) shows a classical circuit with 3 input bits and 1 output bit. The output bit is the result of a boolean function that describes the classical circuit defined in this case by:

$$f(c_1, c_2, c_3) = (c_1 \oplus c_2) \vee (c_1 \wedge c_3) \tag{2.6}$$

Given the value of the output bit and the operations performed in a classical circuit, it still may not be possible to know the values of the input bits. A quantum circuit, on the other hand, as shown in Figure 2.1(b), has exactly as many input qubits as output qubits. In the shown schematic, time moves from left to right, where each line represents the evolution of a single qubit through the sequence of gate cycles in a circuit. The input quantum state, $|\Psi\rangle = |q1, q2, q3\rangle$, is transformed by multiplying $|\Psi\rangle$'s state vector by the 8×8 operator U which describes the evolution of the qubits' state through the smaller suboperators $\{U_1, U_2, U_3, U_4\}$ that make up the circuit.

The operator that describes the first timestep in the circuit is the tensor product of the two matrices, U_1 and U_2, while the operator that describes the last (third) time-step is the tensor product of two identity matrices that leave qubits $q1$ and $q2$ unchanged and the one-qubit U_4 matrix[1]. For two matrices, U_1 and U_2, the tensor product becomes:

$$U_1 \otimes U_2 \ = \ \begin{bmatrix} u_1^{(11)} U_2 & u_1^{(12)} U_2 & \cdots \\ u_1^{(21)} U_2 & u_1^{(22)} U_2 & \cdots \\ \vdots & \ddots & \cdots \end{bmatrix}, \tag{2.7}$$

which is an 8×8 matrix that describes the first timestep. Thus, the final state of $|\Psi\rangle$ after the circuit completes is given by:

$$|\Psi\rangle \ \rightarrow \ U|\Psi\rangle \ = \ [(I \otimes I \otimes U_4) \times (U_3 \otimes I) \times (U_1 \otimes U_2)]\,|\Psi\rangle, \tag{2.8}$$

where the state $|\Psi\rangle$ is first multiplied by $(U_1 \otimes U_2)$, then by $(U_3 \otimes I)$, and finally by $(I \otimes I \otimes U_4)$. The one-qubit matrix denoted by the letter I is the 2×2 identity matrix that does nothing on a single qubit

$$I \begin{bmatrix} a \\ b \end{bmatrix} \ = \ \begin{bmatrix} 1 & 0 \\ 0 & 1 \end{bmatrix} \begin{bmatrix} a \\ b \end{bmatrix} \ = \ \begin{bmatrix} a \\ b \end{bmatrix} \tag{2.9}$$

Given the final state of a quantum circuit, applying the inverse of the operations in reverse will bring the state back to its original form.

---

[1]The tensor product is an operation denoted by the symbol "$\otimes$" between two matrices where each element of the first matrix is replaced by the second matrix, scaled by that element.

Similar to how the classical $NAND$ gate is a universal gate for classical computation, a logically universal set of unitary operations for quantum computing is given by the one and two-qubit gates:

$$H = \frac{1}{\sqrt{2}}\begin{bmatrix} 1 & 1 \\ 1 & -1 \end{bmatrix}, \quad \Phi_\phi = \begin{bmatrix} 1 & 0 \\ 0 & e^{i\phi} \end{bmatrix}, \quad cnot = \begin{bmatrix} 1 & 0 & 0 & 0 \\ 0 & 1 & 0 & 0 \\ 0 & 0 & 0 & 1 \\ 0 & 0 & 1 & 0 \end{bmatrix}, \quad (2.10)$$

where any $2^n \times 2^n$ unitary matrix can be composed of the three matrices shown above. The first gate in Equation 2.10 is the one-qubit Hadamard gate, denoted with the letter H. The Hadamard gate takes the state $|0\rangle$ to the new state marked as $|+\rangle$, and the state $|1\rangle$ to the new state marked as $|-\rangle$. Each of the two states, $|+\rangle$ and $|-\rangle$, is simply an equal superposition of the states $|0\rangle$ and $|1\rangle$ and are defined as:

$$|+\rangle = H|0\rangle = \frac{1}{\sqrt{2}}|0\rangle + \frac{1}{\sqrt{2}}|1\rangle = \frac{1}{\sqrt{2}}(|0\rangle + |1\rangle) \quad (2.11)$$

$$|-\rangle = H|1\rangle = \frac{1}{\sqrt{2}}|0\rangle - \frac{1}{\sqrt{2}}|1\rangle = \frac{1}{\sqrt{2}}(|0\rangle - |1\rangle), \quad (2.12)$$

The phase gate Φ_ϕ leaves the $|0\rangle$ element unchanged but applies a rotation of ϕ radians to the $|1\rangle$ state by multiplying it by the quantity $e^{i\phi}$. The Hadamard gate and the Φ_ϕ gate form a universal set of single-qubit gates, where any valid 2×2 unitary matrix can be approximated by these two gates. One can verify that multiplying any arbitrary two-qubit vector $[a, b]^T$ that describes the state of a single qubit by the matrix for the Φ_ϕ operator will result in a two-element vector with the a coefficient unchanged, while the b coefficient will be multiplied by a factor of $e^{i\phi}$. Finally, the two-qubit *controlled-NOT* (i.e., CNOT) gate makes the interaction between any two-qubits possible. The CNOT gate flips the state of the target qubit whenever the control qubit is set. Its action on an arbitrary two-qubit state, $|\Psi\rangle = (c_0|00\rangle + c_1|01\rangle + c_2|10\rangle + c_3|11\rangle)$, described by the vector $[c_0, c_1, c_2, c_3]^T$ is:

$$U_{cnot}|\Psi\rangle = \begin{bmatrix} 1 & 0 & 0 & 0 \\ 0 & 1 & 0 & 0 \\ 0 & 0 & 0 & 1 \\ 0 & 0 & 1 & 0 \end{bmatrix} \begin{bmatrix} c_0 \\ c_1 \\ c_2 \\ c_3 \end{bmatrix} = \begin{bmatrix} c_0 \\ c_1 \\ c_3 \\ c_2 \end{bmatrix}, \quad (2.13)$$

where the last two elements of the state vector have been flipped. The effect of the CNOT gate, where the first qubit is control and second is target, on the state $|ab\rangle$ is the new state $|a(a \oplus b)\rangle$:

$$|00\rangle \rightarrow |00\rangle; \quad |01\rangle \rightarrow |01\rangle; \quad |10\rangle \rightarrow |11\rangle; \quad |11\rangle \rightarrow |10\rangle$$

Any n-qubit unitary transformation can be decomposed into a combination of only CNOT, Hadamard, and the phase Φ_ϕ gates, where the phase angle need only be $\phi = \pi/2$ or $\phi = \pi/4$ radians. The phase gates with angles of $\pi/2$ and $\pi/4$ radians are known as the S and T gates, respectively:

$$S = \begin{bmatrix} 1 & 0 \\ 0 & e^{i\frac{\pi}{2}} \end{bmatrix} = \begin{bmatrix} 1 & 0 \\ 0 & i \end{bmatrix}, \quad T = \begin{bmatrix} 1 & 0 \\ 0 & e^{i\frac{\pi}{4}} \end{bmatrix} \tag{2.14}$$

An important set of single-qubit gates, known as the *Pauli* matrices, are the four gates shown below denoted with the letters $\{I, X, Y, Z\}$:

$$I = \begin{bmatrix} 1 & 0 \\ 0 & 1 \end{bmatrix}, \quad X = \begin{bmatrix} 0 & 1 \\ 1 & 0 \end{bmatrix}, \quad Z = \begin{bmatrix} 1 & 0 \\ 0 & -1 \end{bmatrix}, \quad Y = -iZX = \begin{bmatrix} 0 & -i \\ i & 0 \end{bmatrix}, \tag{2.15}$$

The X gate is the *bit-flip* gate, which takes the state $|0\rangle$ to $|1\rangle$ and $|1\rangle$ to $|0\rangle$. The CNOT gate, defined in Equation 2.10 is nothing more than a controlled-X gate. The Z gate is a 180 degree rotation of the phase (known as the *phase-flip*) gate which leaves $|0\rangle$ unchanged and takes $|1\rangle$ to $-|1\rangle$. Note that the Z gate is the phase gate with the angle $\phi = \pi$ radians, while the X gate can be constructed by conjugating the Z matrix with the Hadamard matrix: $X = HZH$. The Y gate can be obtained by multiplying the X and Z gates together with a global phase factor of $-i$ (i.e., $Y = -iZX$).

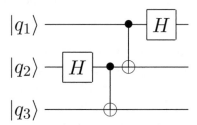

Figure 2.2: Example circuit consisting of two Hadamard gates and two CNOT gates.

To show how the state of a quantum system evolves through the application quantum gates, we consider the the three-qubit circuit example shown in Figure 2.2. The example has two CNOT gates and two Hadamard gates. The CNOT gate's circuit representation is usually drawn with an $\oplus$ symbol to mark the fact that the gate's function is to perform the *xor* operation between the control qubit and the target qubit if the control qubit is set.

Suppose now that the input state of the first qubit in Figure 2.2 is an arbitrary qubit state $|q_1\rangle = a|0\rangle + b|1\rangle$ and the other two qubits, $|q_2\rangle$ and $|q_3\rangle$, are both initialized to $|0\rangle$. Thus, before the first timestep the state of the entire system is

$$|q_1 q_2 q_3\rangle = (a|0\rangle + b|1\rangle))|00\rangle = a|000\rangle + b|100\rangle, \tag{2.16}$$

where the i'th entry in each bitstring state $|x_n x_{n-1} \cdots x_i \cdots x_0\rangle$ denotes the state of the i'th qubit. The 8-element probability amplitude vector that describes the state of the system has all zero entries except the zeroth entry (equal to a) and the fourth entry, which is equal to b.

Following the first Hadamard gate on the second qubit, the state of the system becomes:

$$\frac{1}{\sqrt{2}}(a|000\rangle + a|010\rangle + b|100\rangle + b|110\rangle) \tag{2.17}$$

The first CNOT gate flips the state of qubit $|q_3\rangle$ where ever $|q_3\rangle$ is $|1\rangle$, so the state of the three qubits after the two CNOT gates becomes $\frac{1}{\sqrt{2}}(a|000\rangle + a|011\rangle + b|110\rangle + b|101\rangle)$. Finally, the application of the Hadamard gate on qubit $|q_1\rangle$ places the final state of the three qubits into the superposition:

$$|q_1 q_2 q_3\rangle = \frac{1}{2}(a(|000\rangle + |011\rangle + |100\rangle + |111\rangle) + b(|010\rangle + |001\rangle - |110\rangle - |101\rangle)), \tag{2.18}$$

where we have factored out common terms, the probability coefficients a and b. The global phase factor of $\frac{1}{2}$, introduced by the successive application of the two H gates, can be left out, since it doesn't functionally change the probability values for the coefficients for each state relative to the other states. Also, the coefficients are phase factors that can be moved to any location within their corresponding state. For example, $a|00\rangle$ can be written as $|0\rangle a|0\rangle$. Thus, we can rewrite the final state of our example circuit by factoring out some common terms and moving the coefficients around to get:

$$|00\rangle(a|0\rangle + b|1\rangle) \ + \ |01\rangle(b|0\rangle + a|1\rangle) \ + \ |10\rangle(a|0\rangle - b|1\rangle) \ + \ |11\rangle(a|1\rangle - b|0\rangle) \tag{2.19}$$

Note that the state of qubit $|q_3\rangle$ is any of four different arbitrary qubit states that look very much like the initial state of qubit $|q_1\rangle$. Thus, the state of qubit $|q_1\rangle$ has been recreated in qubit $|q_3\rangle$ (with some error) without directly interacting the qubits. The error depends on the measurement results when measuring qubits $|q_1\rangle$ and $|q_2\rangle$. In the next section we will see that measuring the values of qubits $|q_1\rangle$ and $|q_2\rangle$ tells us what the error is and this information can be used to correct the error.

2.3   QUANTUM MEASUREMENT

Reading values from a classical register is a trivial operation. Values can be read reliably and copied to other registers. Unfortunately, this is not the case for quantum registers. Reading out the state of any qubit of a quantum register involves a measurement that destroys the superposition of that qubit, effectively terminating any quantum computation which requires the state of the qubit prior to measurement.

Furthermore, just as there are many unitary operators U that can be applied to a single qubit, there are many ways to perform a measurement on a qubit. A measurement of a single qubit can be performed along the eigenbasis of any one-qubit operator U. The result of the measurement is a classical number which corresponds to one of U's eigenvalues, and the effect of the measurement operation on the state of the qubit is to collapse the qubit's state into one of U's eigenvectors

Consider, for example, measurement along the eigenbasis of the Z gate, also known as measurement along the *computational basis* [2] since Z's eigenvectors are the states:

$$|0\rangle = \begin{bmatrix} 1 \\ 0 \end{bmatrix}, \quad |1\rangle = \begin{bmatrix} 0 \\ 1 \end{bmatrix}$$

with eigenvalues 1 and -1, respectively. If the result of performing a Z measurement is 1, then the qubit state is collapsed into the state $|0\rangle$ and the result is denoted with the classical bit "0". If, on the other hand, the result of the measurement is -1, the state would be collapsed into the state $|1\rangle$ and result would be denoted as the classical bit "1". For a general qubit state $|\Psi\rangle = a|0\rangle + b|1\rangle$, the result of the measurement will be "0" with probability $|a|^2$ and "1" with probability $|b|^2$. Another option (if allowed by the underlying technology) is measurement along the eigenbasis of the X operator with eigenvectors given by $|+\rangle$ and $|-\rangle$, where:

$$|+\rangle = \frac{1}{\sqrt{2}}(|0\rangle + |1\rangle) = \frac{1}{\sqrt{2}}\left(\begin{bmatrix} 1 \\ 0 \end{bmatrix} + \begin{bmatrix} 0 \\ 1 \end{bmatrix}\right) = \frac{1}{\sqrt{2}}\begin{bmatrix} 1 \\ 1 \end{bmatrix}$$

(2.20)

$$|-\rangle = \frac{1}{\sqrt{2}}(|0\rangle - |1\rangle) = \frac{1}{\sqrt{2}}\left(\begin{bmatrix} 1 \\ 0 \end{bmatrix} - \begin{bmatrix} 0 \\ 1 \end{bmatrix}\right) = \frac{1}{\sqrt{2}}\begin{bmatrix} 1 \\ -1 \end{bmatrix},$$

(2.21)

The general single qubit state $a|0\rangle + b|1\rangle$ can be written in terms of the states $|+\rangle$ and $|-\rangle$ as:

$$\frac{a+b}{\sqrt{2}}|+\rangle + \frac{a-b}{\sqrt{2}}|-\rangle$$

(2.22)

Thus, when measuring the qubit in the X eigenbasis, the resulting state would be collapsed into the state $|+\rangle$ or the state $|-\rangle$ with probabilities equal to $\left|\frac{a+b}{\sqrt{2}}\right|^2$ and $\left|\frac{a-b}{\sqrt{2}}\right|^2$, respectively. For simplicity, we will consider measurement in the computational basis *only* and will also represent n-qubit states in the computational basis.

The destructive nature of measurement is one of the reasons why quantum states cannot be copied, as shown by the quantum no-cloning theorem given by Zurec [225]. The no-cloning theorem means that FANOUT is impossible in quantum computing, which has significant implications on the methods for communicating quantum information across a computing device and the efficiency of quantum data communication. We cannot, for example, simply transmit quantum information on a wire to a different destination while leaving the source unmodified. We must design the system such that qubits or the states of the qubits are physically moved between two locations.

---

[2]Up to this point, we have described qubit states in the computational basis, i.e., $|\Psi\rangle = a|0\rangle + b|1\rangle$.

2.4   EXAMPLE: THE 3-QUBIT QUANTUM TOFFOLI GATE

The Toffoli gate in classical computation is defined as the *controlled-controlled-NOT* gate, which flips the state of the target bit if the states of both control bits are set: $(a, b, c) \rightarrow (a, b, c \oplus ab)$. In classical computation, the Toffoli gate is particularly important because it is the smallest universal, reversible classical operation [205]. It is universal, because it can simulate the NAND gate if the third bit is fixed to 1 at the input. It is reversible, because applying the Toffoli gate again will bring the state of the three bits back to (a, b, c). Quantum mechanically, the Toffoli gate takes the state $|abc\rangle$ to the state $|ab(c \oplus ab)\rangle$ as follows:

$$|000\rangle \rightarrow |000\rangle; \quad |001\rangle \rightarrow |001\rangle; \quad |010\rangle \rightarrow |010\rangle; \quad |011\rangle \rightarrow |011\rangle$$
$$|100\rangle \rightarrow |100\rangle; \quad |101\rangle \rightarrow |101\rangle; \quad |110\rangle \rightarrow |111\rangle; \quad |111\rangle \rightarrow |110\rangle$$

The circuit representation for the Toffoli gate is shown in Figure 2.3, along with the gate's 8×8 unitary matrix.

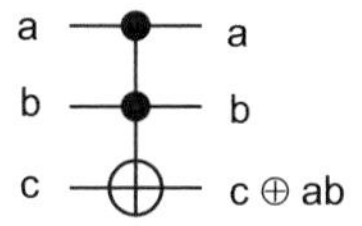

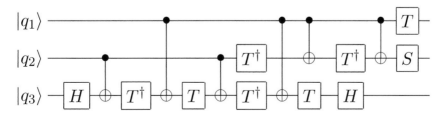

$$U_{toffoli} = \begin{pmatrix} 1 & 0 & 0 & 0 & 0 & 0 & 0 & 0 \\ 0 & 1 & 0 & 0 & 0 & 0 & 0 & 0 \\ 0 & 0 & 1 & 0 & 0 & 0 & 0 & 0 \\ 0 & 0 & 0 & 1 & 0 & 0 & 0 & 0 \\ 0 & 0 & 0 & 0 & 1 & 0 & 0 & 0 \\ 0 & 0 & 0 & 0 & 0 & 1 & 0 & 0 \\ 0 & 0 & 0 & 0 & 0 & 0 & 0 & 1 \\ 0 & 0 & 0 & 0 & 0 & 0 & 1 & 0 \end{pmatrix}$$

Figure 2.3: Circuit representation and the three-qubit 8×8 matrix for the Toffoli gate.

A circuit that implements the Toffoli composed of just Hadamard, T, S, and CNOT gates is shown in Figure 2.4. $T^\dagger$ is the complex conjugate of the matrix that implements the T gate. The ability to implement the Toffoli gate in one and two-qubit gates is important, since physical gates for known device technologies only allow one and two-qubit operations.

Figure 2.4: Circuit implementation of the Toffoli gate using only one and two-qubit gates.

The Toffoli gate is very important in the implementation of many quantum algorithms. It is central, for example, in the implementation of quantum adders. The implementation of a reversible 2-bit adder using Toffoli and CNOT gates is shown in Figure 2.5. The circuit adds the two bitstrings "(x_1, x_2)" and "(s_1, s_2)", where the least significant bit is the leftmost bit. The result is stored in the bitstring "(s_1, s_2, c), where "c" is the carry-out bit. For example, the addition of the input strings "(1, 1)" and "(1, 1)" should yield the result "$(s_1 = 0, s_2 = 1, c = 1)$". An additional ancillary bit, stored in the state of qubit $|a\rangle$ is used. If the information were stored in qubits, the circuit in Figure 2.5 becomes a quantum 2-bit adder, based on the classical ripple-carry adder. If the input is a superposition of all possible combinations for the input strings, then the output would be a superposition, where each state holds the result of the addition. Quantum adders are integral to the circuit for quantum modular exponentiation used in Shor's quantum factoring algorithm.

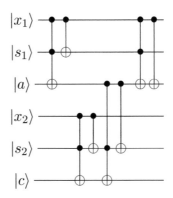

Figure 2.5: Two-Bit Adder composed of quantum controlled-NOT (CNOT) and Toffoli gates. The circuit adds the two bitstrings "(x_1, x_2)" and "(s_1, s_2)", where the least significant bit is the leftmost bit. The result is stored in the bitstring "(s_1, s_2, C), where C is the carry-out bit.

2.5   EXAMPLE: QUANTUM FOURIER TRANSFORM (QFT)

Another example is the circuit for the quantum fourier transform (QFT), which is integral to many quantum algorithms. An N-qubit QFT maps a quantum state $|\Psi\rangle$ placed into a superposition of all states $|x_i\rangle$, where each x_i is an integer less than N, into the *Fourier basis*. The Fourier basis is a superposition of the quantum states $|\chi_a\rangle$ defined by:

$$|\chi_a\rangle = \frac{1}{\sqrt{N}} \sum_{j=0}^{N-1} e^{2\pi i \frac{aj}{N}} |j\rangle \tag{2.23}$$

The corresponding N-qubit QFT circuit, which performs the mapping of $|\Psi\rangle$ into the Fourier basis, consists of Hadamard gates and controlled-phase $Rz(\theta)$ rotations where $\theta = \pi/2^d$ radians, instead d is an integer such that $1 \leq d \leq (N-1)$. The $Rz(\theta)$ gate is a rotation around the $\hat{z}$-axis and is defined by the unitary operator $e^{-i\theta Z}$, where Z is the single-qubit Pauli Z gate.

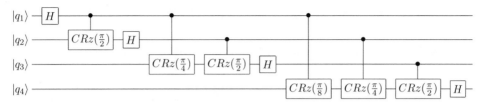

Figure 2.6: Circuit for the 4-qubit Quantum Fourier Transform.

The circuit for a 4-qubit QFT is shown in Figure 2.6. It can be seen that the implementation of the QFT is efficient on a quantum computer and requires $\mathcal{O}(N^2)$ gates [56]. The caveat is that the implementation of the controlled-phase $Rz(\theta)$ gates is non-trivial. Song and Klappenecker [189] have shown that an arbitrary two-qubit controlled operator can be implemented using, at most, two CNOT gates and three single-qubit gates.

2.6   EXAMPLE: QUANTUM TELEPORTATION

Another important quantum operation is quantum teleportation [21]. Quantum teleportation allows us to transfer the state of one qubit in location A to another qubit in location B without the need for direct interaction between the two qubits. Quantum teleportation is also a universal primitive for quantum computation [87], much like the three gates CNOT H, and Phase, shown in Equation 2.10, are a universal set of gates.

If we look back at the example of Figure 2.2, we see that upon measurement of both qubits, $|q_1\rangle$ and $|q_1\rangle$, we will obtain one of the four strings "00", "01", "10", or "11". If the result is "00", the state of qubit $|q_3\rangle$ after measurement will be $a|0\rangle + b|1\rangle$ (see Equation 2.19, which is the original state of qubit $|q_1\rangle$. Thus, upon observation of the string "00", the initial value of qubit $|q_1\rangle$ has been *teleported* to qubit $|q_3\rangle$ without having to interact with qubits $|q_1\rangle$ and $|q_3\rangle$ *directly*.

Even if the result is any of the other strings "01", "10", or "11", one can see from Equation 2.19 that the initial state of qubit $|q_1\rangle$ is recreated in the state of $|q_3\rangle$, with some error. The error can be corrected by applying a combination of one bit-flip X gate and one phase-flip Z gate, depending on the measurement result. The full teleportation circuit, complete with the two measurement operations on qubits $|q_1\rangle$ and $|q_2\rangle$ and the recovery X and Z operations on qubit $|q_3\rangle$, is shown in Figure 2.7. Teleportation does not violate the no cloning theorem, since the state of qubit $|q_1\rangle$ is teleported into the state of $|q_3\rangle$ only after it is destroyed by the measurement operation.

2.7   EXAMPLE: DEUTSCH'S QUANTUM ALGORITHM

The destructive nature of quantum measurement raises an important question about the limits of quantum computing and, more specifically, the limits of quantum parallelism. Even though a quan-

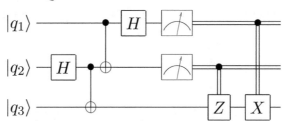

Figure 2.7: Complete circuit for quantum teleportation. The first 4 gates are the same as the ones shown in Figure 2.2. The measurement operations and the recovery operation of X and Z gates are added to complete the teleportation of qubit $|q_1\rangle$ into the state of qubit $|q_3\rangle$, without direct interaction between the two qubits.

tum computer can be used to evaluate a function $f(x)$ over all possible inputs x in a single operation and store the solutions into the state of the quantum computer's internal qubit registers, the quantum computer cannot be used to return the solution of $f(x)$ for a specific input x. Instead, quantum computation can be used to calculate efficiently some global property of $f(x)$ that depends on multiple inputs to the functions. Such global properties include the period of a function and the function's minima, maxima. Deutsch's algorithm [64, 66] is perhaps the simplest quantum algorithm that demonstrates how quantum parallelism can be harnessed to calculate global properties of functions more efficiently that classical algorithms.

Before we describe Deutsch's Algorithm, let us consider how quantum parallelism works when evaluating $f(x)$ as described in Reference [149]. Suppose we start with two qubits in the initial state $|\Psi\rangle = |00\rangle$, and the two-qubit unitary transformation U_f which takes the state $|ab\rangle$ to $|a, b \oplus f(a)\rangle$. The transformation U_f can be simply the CNOT gate, where $f(a) = 0$ if $a = 0$ and $f(a) = 1$ if $a = 1$. Applying the Hadamard gate on the first qubit we obtain the state:

$$|\Psi\rangle \quad \rightarrow \quad H_1|\Psi\rangle \quad = \quad \frac{1}{\sqrt{2}}(|00\rangle + |10\rangle) \tag{2.24}$$

The Hadamard gate is the key to accessing quantum parallelism, as it transforms any state $|a\rangle$ into a superposition of all possible values of a (namely "0" and "1"). After the unitary transformation U_f on the two qubit state $|\Psi\rangle$, the state takes the form:

$$|\Psi\rangle \quad \rightarrow \quad U_f|\Psi\rangle \quad = \quad \frac{1}{\sqrt{2}}(|0, \ f(0)\rangle + |1, \ f(1)\rangle) \tag{2.25}$$

Note that the state of $|\Psi\rangle$ contains the evaluated function $f(x)$ for both possible inputs "0" and "1", which we have derived with a single clockstep. The problem is that, upon measurement of one of the two qubits, we will obtain information about the function $f(x)$ for only one input.

A simple version of Deutsch's algorithm [149] is similar to the above example, but instead, we start with the two-qubit state $|\Psi\rangle = |01\rangle$ and apply a Hadamard gate to both qubits before we apply the unitary transformation U_f. After the two Hadamard gates, we have the state:

$$|\Psi\rangle \rightarrow H_1 H_2 |\Psi\rangle = \frac{1}{\sqrt{2}}(|0\rangle + |1\rangle) \otimes \frac{1}{\sqrt{2}}(|0\rangle - |1\rangle) \tag{2.26}$$

If we apply another Hadamard gate on the first qubit *after* the two-qubit U_f transformation, we obtain the following final state:

$$|\Psi\rangle_{final} = \pm |f(0) \oplus f(1)\rangle \left[\frac{|0\rangle - |1\rangle}{\sqrt{2}} \right] \tag{2.27}$$

In Equation 2.27, we see the state of the first qubit is the quantity $f(0) \oplus f(1)$ for the function $f(x)$, which is a global result for the function $f(x)$ that depends on both inputs. Unlike probabilistic classical computation where the two alternatives of $f(x)$ exclude one another, quantum states that can be used represent classical bitstrings which can actually interfere with each other. The interference was induced by the third Hadamard gate, which was applied to the first qubit in Deutsch's algorithm.

In general, the design of quantum algorithms involves the identification of some global property of a function $f(x)$ over the function's inputs and the design of a quantum circuit that allows a quantum computer to efficiently compute this global property. One example is using the Fourier transform to force a quantum state into a superposition such that all superposition states, with values corresponding to the period of some periodic function have a higher probability of being measured [187]. The calculated period is subsequently used to find the factors of a large number N in Shor's factoring algorithm [184].

2.8   QUANTUM ENTANGLEMENT AND EPR PAIRS

Deutsch's algorithm (and generally the ability to utilize quantum parallelism) works because the states of multiple quantum systems, (i.e., qubits) can be physically *entangled* with one another. This means the probability amplitudes that describe the state of qubit $|q\rangle$ are changed when another qubit $|p\rangle$ is modified or measured whenever qubits $|q\rangle$ and $|p\rangle$ are entangled. Consider, for example, equation 2.19, where qubits $|q_1\rangle$ and $|q_2\rangle$ are entangled with the state of qubit $|q_3\rangle$. When measuring qubits $|q_1\rangle$ and $|q_2\rangle$ their states are collapsed as a state that corresponds to the result of the measurement. However, the state of qubit $|q_3\rangle$ is also collapsed accordingly. This dependence of amplitudes between different qubits is known as *quantum entanglement*, which together with the existence of quantum states in a superposition, is the most powerful and least understood behavior of quantum mechanical physical systems.

Quantum entanglement is the driving engine behind all quantum algorithm implementations, including quantum teleportation, quantum key distribution, superdense coding [25, 92], and even quantum error correction. Independently, each qubit is its own entity since there is some probability of obtaining either a $|0\rangle$ or a $|1\rangle$ when measuring any qubit by itself. However, when multiple qubits are entangled, any action (such as a gate or a measurement) on a single qubit will affect the states of all qubits entangled with it. Entanglement is also the primary reason why quantum computers are so difficult to physically realize and is the principle source for quantum decoherence: the entanglement of the quantum system with the surrounding environment and the resulting loss of the system's desired superposition state.

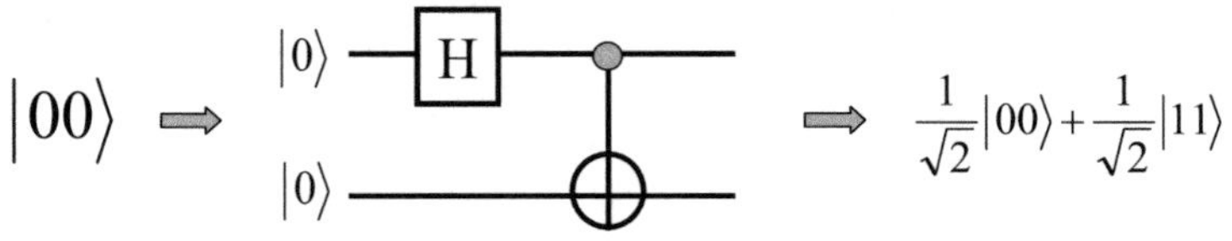

$$|00\rangle \implies \quad \begin{array}{c} |0\rangle - \boxed{H} - \bullet - \\ |0\rangle - \oplus - \end{array} \quad \implies \quad \frac{1}{\sqrt{2}}|00\rangle + \frac{1}{\sqrt{2}}|11\rangle$$

Figure 2.8: Creation of a maximally entangled EPR pair.

One very important entangling gate between two qubits is the CNOT gate. Consider Figure 2.8, where we begin with two qubits initialized to the state $|00\rangle$. The application of a Hadamard gate on the first qubit sends the system into the state: $(|0\rangle + |1\rangle)|0\rangle$, which can be rewritten as $|00\rangle + |10\rangle$, where the first qubit is in an equal superposition of $|0\rangle$ and $|1\rangle$ and the second qubit remains $|0\rangle$. A CNOT gate with the first qubit as target flips the state of the second qubit only when the first qubit is $|1\rangle$, giving us the entangled state $(|00\rangle + |11\rangle)/\sqrt{2}$. This fully entangled state is known as an EPR pair, named after its discoverers in 1935, Einstein, Podolsky, and Rosen. The two qubits are entangled in the sense that whenever the first qubit is measured and the result is the bit "0", then not only is the state of the first qubit destroyed, but the state of the second qubit, would also be collapsed to $|0\rangle$.

EPR pairs are also known as two-qubit *cat states*. An *n*-qubit cat state can be generalized to $|\Psi\rangle = |00...0\rangle + |11...1\rangle$. An analogy for a four-qubit cat state using four cubes, [145, 224] drawn without a particular frame of reference, is shown schematically in the top-most part of Figure 2.9. The moment an observer is shown a *single* cube with the south face brought forward, or the north face brought forward (equivalent to quantum measurement), the observer's mind will be immediately fixed to the shown frame of reference for *all* cubes. Thus, showing an observer a particular frame of reference for one cube is equivalent to measuring not just the shown cube, but the entire entangled set of four cubes. The figure shows the two possible outcomes of observing either frame of reference in the bottom rows of four cubes each.

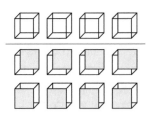

Figure 2.9: Four cubes that aid in visualizing a 4-qubit cat state.

EPR pairs play an integral part in the quantum teleportation protocol described in Figure 2.7. Note that the first Hadamard gate and the first CNOT gate are used to prepare an EPR pair between qubits $q2$ and $q3$, which are then entangled with the data qubit $q1$ through the second CNOT gate in Figure 2.7. In general, quantum teleportation works by interacting an arbitrary data qubit with a previously prepared two-qubit EPR pair, such that the state of the data qubit becomes the state of one of the EPR qubits (up to some error).

2.9   OTHER MODELS OF QUANTUM COMPUTATION

Until now, we have described the circuit model for quantum computation, which allows the execution of algorithms in the form of a sequence of operations applied on a number of two-state quantum systems, called qubits. For the rest of this work, we restrict ourselves to quantum circuits composed of a small set of universal quantum gates. These include any arbitrary single-qubit operations, the CNOT gate, and measurement.

Other examples of computational models (not explored here) are adiabatic quantum computation [5, 48, 74], cluster state quantum computation [37, 148, 150, 227], geometric quantum computation [105], and the theory of topological quantum computation [79]. In adiabatic quantum computation, the computer is initialized with some initial Hamiltonian, H_i. H_i is then adiabatically deformed into a final Hamiltonian, H_f, that represents the solution to the problem being calculated. Cluster states are a collection of highly entangled qubits with the property that arbitrary quantum computation can be performed purely through single-qubit measurement operations. Topological quantum computation uses hypothetical quantum systems with particular kinds of topological excitations to prevent decoherence. Recent studies suggest that such systems may exist in nature [33, 34, 45].

Combined, the variety of quantum computation models provide different methods for extending the application space for quantum computers, and may some day redefine the system design of a large-scale machine. In this book, however, the focus is on the circuit model, which we use to describe a digital, scalable quantum architecture scheme that overcomes the primary scalability issues of size and resource distribution. The circuit model is universal and can be used to efficiently simulate all other models of quantum computation, thus the architecture described is (in theory) a general purpose quantum architecture.

CHAPTER 3

Key Quantum Algorithms

This chapter introduces key quantum algorithms. Understanding the structure and subtleties of the quantum applications we desire to execute on a quantum computer makes it possible to design a system that is tailored to our needs.

The topics discussed in this section can be grouped into the three main catagories shown in the following list. Subsequent subsections introduce the key algorithms pertaining to these topics and illustrate their correctness.

1. **Integer factorization and Hidden Subgroup Problems** Quantum algorithms for polynomial time integer factorization [184] are possibly the most famous and commonly mentioned quantum algorithms, due to their potential for breaking conventional cryptosystems. We will provide an overview of the integer factorization, explanations/proof overviews of its correctness, and also a discussion of hidden subgroup problems (a class of problems in which the integer factorization problem is a special case).

2. **Grover's algorithm for quantum search** Grover's algorithm [89] is another well-known quantum algorithm. Grover's algorithm is an algorithm for determining properties of a function given a quantum black-box representation of the function, in essence, "searching" for a certain element in the domain of the function.

3. **Quantum adiabatic algorithms** Computation in quantum adiabatic algorithms [75] is done by evolving a quantum system whose behavior is governed by the Schrodinger equation. We present an overview of the mathematical model of quantum adiabatic computation as well as an example of a quantum adiabatic algorithm.

The explanations of the first two topics in this chapter closely follow the expositions presented by Kaye et al. [110]. We refer the reader to [110, 149] for more detailed explanations of the derivations and underlying principles behind these algorithms.

3.1 QUANTUM INTEGER FACTORIZATION

3.1.1 THE INTEGER FACTORIZATION PROBLEM

The integer factorization problem can simply be stated as: given a composite N, determine m prime numbers $p_1, \ldots, p_m$ such that $p_1 \times \ldots \times p_m = N$. The case where $m = 2$ (the composite is a multiple of only two primes) is of particular interest, as a composite number with these properties is utilized in the RSA public-key cryptosystem [172], where untrusted entities can encrypt using a public key, yet only entities with a private-key can decrypt this encrypted data. RSA's security relies upon the difficulty of factoring the composite into two primes. If such a feat of factorization is possible (efficiently, of course), then it is possible to decode data encoded using RSA without a private key, thereby breaking the security of the crypto-system. No polynomial time classical algorithm for factoring is currently known, but (as mentioned earlier in this section) techniques utilizing quantum algorithms for doing so have been devised.

3.1.2   A QUANTUM INTEGER FACTORIZATION ALGORITHM

In 1994, Peter Shor's groundbreaking paper "Algorithms for quantum computation: Discrete logarithms and factoring" sent waves through the computer science community by demonstrating a quantum algorithm for integer factorization. Shor's seminal contribution is commonly credited with igniting a wider interest in quantum computation, mainly because of the exciting (and possibly perilous) prospect of an algorithm for solving precisely the problems that many cryptosystems are built upon. In this text, we will look at Shor's original algorithm using the eigenvalue/phase estimation analysis technique described by Kaye et al. in [110]. This analysis is based on the work initially done by Kitaev in [115].

   The actual quantum algorithms used in the "quantum" techniques for integer factorization do not directly produce the factors of a given number. Rather, the quantum algorithms involved solve a related problem of order finding. Order finding is the task of computing the smallest period r of a function $f(x)$ such that $f(x) = f(x + r)$. Solving the order-finding problem with the function $f(x) = a^x \mod N$ for some $a \in \mathbb{Z}, a < N$ allows us to solve the integer factorization problem. Quantum integer factorization algorithms usually look something like the following:

1. **Given the input information (namely, a large composite), a sequence of efficient classical steps reduces the problem to the order-finding problem.** Specifically, given an integer N, we first choose a random number $a < N$. Next, compute $gcd(a, N)$. If $gcd(a, N) \neq 1$, then a and $gcd(a, N)$ are factors of N and we are done.

2. **A quantum algorithm (augmented by an efficient classical algorithm involving continued fractions) solves the order finding problem (with high-probability).** A quantum algorithm solves the order-finding problem for $f(x) = a^x \mod N$. This algorithm (including its classical component) will be discussed in more detail in later sections.

3. **Another sequence of classical steps efficiently verifies the output of the quantum algorithm.** Here, we check if r is odd or $a^{r/2} = -1 \mod N$. In both cases, we go back to step 1). Otherwise, the factors of N are $gcd(a^{r/2} \pm 1, N)$ and we are done.

Of course, such a procedure raises a number of questions for the discerning reader. Specifically, how do we reduce the integer factorization problem to order-finding and how do we know numbers produced in the last step are the factors of N? The following subsection provides a proof of the classical parts of steps 1 and 3 of this algorithm.

3.1.3 QUANTUM INTEGER FACTORIZATION: PROOF OF CLASSICAL PART

In the first step of the factorization algorithm, we selected a random integer a and we want to find the minimum period r of the function $f(x) = a^x \mod N$. Since r is the period $f(x) = a^x \mod N$, we have

$$a^r \equiv 1 \mod N \tag{3.1}$$

If r is even (which it will be, by step 3), this becomes

$$
\begin{aligned}
a^r &\equiv 1 \mod N & (3.2)\\
a^r - 1 &\equiv 0 \mod N & (3.3)\\
(a^{r/2} - 1)(a^{r/2} + 1) &\equiv 0 \mod N. & (3.4)
\end{aligned}
$$

Let $a_0 = a^{r/2} - 1$ and $a_1 = a^{r/2} + 1$. This implies that $N|(a^{r/2} - 1)(a^{r/2} + 1)$ or $N|a_0 a_1$. Note that in the algorithm

- Since r is the smallest integer such that $a^r = 1$, N cannot divide a_0.

- We ensure that N does not divide a_1, as we check this and start over if this is the case.

If we can show that if N does not divide either a_0 or a_1, then N has a non-trivial factor in common with each of a_0 and a_1. This would imply that N has factors $gcd(N, a^{r/2} - 1)$ and $gcd(N, a^{r/2} + 1)$. The following is a proof (by contradiction) for this. First, assume that N does not divide either a_0 or a_1 ($gcd(N, a_0) = gcd(N, a_1) = 1$). Since $N|a_0 a_1$ (as shown earlier), we have $kN = a_0 a_1$ for some integer k. This would imply that there exist integers m, n such that $ma_0 + nN = 1$. But note that

$$
\begin{aligned}
ma_0 + nN &= 1 & (3.5)\\
ma_1 a_0 + nNa_1 &= a_1 & (3.6)\\
mkN + na_1 N &= a_1, & (3.7)
\end{aligned}
$$

which implies $N|a_1$, which contradicts our initial assumption.

3.2 ORDER FINDING

Finding the order of the function $f(x)$ described earlier can be solved via a quantum algorithm followed by a classical algorithm involving continued fractions. Figure 3.1 shows a high-level overview of this algorithm, consisting of a quantum circuit for such an algorithm followed by the necessary classical steps. The circuit consists of a quantum Fourier transform, a controlled operator that implements a function (this step is usually referred to as modular exponentiation), and finally an inverse quantum Fourier transform.

As described by Kaye et al. in [110], we can conceptually understand this quantum algorithm in the context of two other problems for which we have quantum algorithms: *eigenvalue estimation* and *phase estimation*. Specifically,

1. Order finding can be framed as an application of eigenvalue estimation.

2. Eigenvalue estimation can be framed as an application of phase estimation

3. Phase estimation can be framed as an application of the inverse quantum Fourier transform.

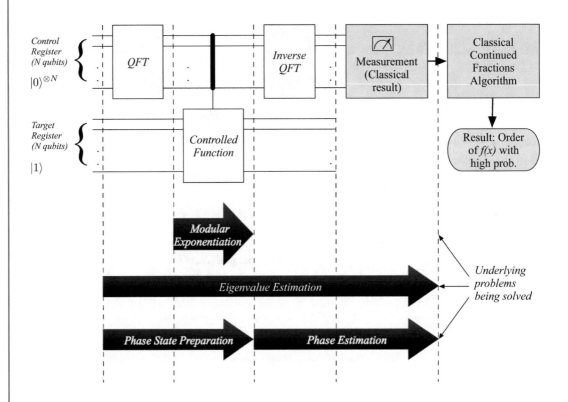

Figure 3.1: High-level overview of the order-finding algorithm.

Figure 3.1 also illustrates the underlying relationship between the stages of the quantum circuit and phase and eigenvalues estimation problems. The phase estimation analysis was first introduced by [115] soon after Shor's initial paper. While it is not necessary to analyze the algorithm in this way (Shor's paper does so in a different manner from Kitaev's), doing so introduces useful primitives that can be used in other algorithms.

The following two subsections will demonstrate the main ideas behind the reduction of order finding to eigenvalue estimation and the classical continued fractions algorithm required to complete the algorithm, respectively. To begin to explain these concepts, we first need to present the phase and eigenvalues estimation problems (subsequent sections of this text will explain these algorithms in more detail):

- Given the quantum state

$$|\psi\rangle = \frac{1}{2^n} \sum_{y=0}^{2^n-1} e^{2\pi i\omega y}|y\rangle, \tag{3.8}$$

the **phase estimation problem** is the problem of finding a good estimate of the phase parameter ω. This process of phase estimation is done via the inverse quantum Fourier transform.

- Given a quantum operator U with eigenstate $|\psi\rangle$ with eigenvalue $e^{2\pi i\omega}$, the **eigenvalue estimation problem** is the problem of attaining a good estimate of ω. In the quantum algorithm for eigenvalue estimation, we first prepare a control register (with value $|0\rangle^{\otimes n}$) and a target register with the eigenstate in question (here $|\psi\rangle$). The algorithm performs the map $|0\rangle^{\otimes n}|\psi\rangle \rightarrow |\tilde{\omega}\rangle|\psi\rangle$; we can now measure the first register to attain an estimate $\tilde{\omega}$ of ω. It is important to note that the eigenvalue estimation process can be viewed as a two stage algorithm with the following stages:

  1. A state preparation stage which creates a state where the eigenvalue in question is a phase term.

  2. An application of the phase estimation algorithm to determine the phase term in the state that was just created, thereby determining the eigenvalue that we want.

3.2.1 ORDER FINDING AS QUANTUM EIGENVALUE ESTIMATION

Given efficient algorithms for solving the eigenvalue and phase estimation problems, we can proceed to derive the order finding procedure. Consider an operator U_a defined as

$$U_a : |s\rangle \rightarrow |sa \mod N\rangle, \tag{3.9}$$

where $0 \leq s < N$. Note that the state

$$|u_k\rangle = \frac{1}{\sqrt{r}} \sum_{s=0}^{r-1} e^{-2\pi i \frac{k}{r} s}|a^s \mod N\rangle \tag{3.10}$$

is an eigenstate of U_a with corresponding eigenvalue $e^{2\pi i \frac{k}{r} s}$. This can be shown as follows:

$$U_a |u_k\rangle = U_a \left[\frac{1}{\sqrt{r}} \sum_{s=0}^{r-1} e^{-2\pi i \frac{k}{r} s} |a^s \mod N\rangle \right] \tag{3.11}$$

$$= \frac{1}{\sqrt{r}} \sum_{s=0}^{r-1} e^{-2\pi i \frac{k}{r} s} U_a |a^s \mod N\rangle \tag{3.12}$$

$$= \frac{1}{\sqrt{r}} \sum_{s=0}^{r-1} \left[e^{2\pi i \frac{k}{r}} e^{-2\pi i \frac{k}{r}} \right] e^{-2\pi i \frac{k}{r} s} |a^{s+1} \mod N\rangle \tag{3.13}$$

$$= e^{2\pi i \frac{k}{r}} \frac{1}{\sqrt{r}} \sum_{s=0}^{r-1} e^{-2\pi i \frac{k}{r}(s+1)} |a^{s+1} \mod N\rangle. \tag{3.14}$$

We can simplify the sum in the following equation as

$$\sum_{s=0}^{r-1} e^{-2\pi i \frac{k}{r}(s+1)} |a^{s+1} \mod N\rangle \tag{3.15}$$

$$= \sum_{s=1}^{r-1} e^{-2\pi i \frac{k}{r} s} |a^s \mod N\rangle + e^{-2\pi i \frac{k}{r} r} |a^r \mod N\rangle \tag{3.16}$$

$$= \sum_{s=1}^{r-1} e^{-2\pi i \frac{k}{r} s} |a^s \mod N\rangle + e^{-2\pi i \frac{k}{r} 0} |a^0 \mod N\rangle \tag{3.17}$$

$$= \sum_{s=0}^{r-1} e^{-2\pi i \frac{k}{r} s} |a^s \mod N\rangle. \tag{3.18}$$

Hence, we have

$$U_a |u_k\rangle = e^{2\pi i \frac{k}{r}} \frac{1}{\sqrt{r}} \sum_{s=0}^{r-1} e^{-2\pi i \frac{k}{r}(s+1)} |a^{s+1} \mod N\rangle \tag{3.19}$$

$$= e^{2\pi i \frac{k}{r}} |u_k\rangle, \tag{3.20}$$

which shows that u_k is an eigenstate of U_a.

If we can compute the eigenvalue $e^{2\pi i \frac{k}{r}}$ of an eigenstate u_k, we can then determine r, which then allows us to compute the period via a continued fractions algorithm (as shown in the next subsection). We can use the eigenvalue estimation algorithm to get this eigenvalue. We cannot simply apply the eigenvalue estimation algorithm to a random state $|0\rangle |u_k\rangle$ for $0 \le k < r - 1$, which will map to the state $|\tilde{k}/r\rangle |u_k\rangle$, as we need to know r in order to prepare this. However (as we will see in moment), we do not need to know r in order to apply the algorithm to $\frac{1}{\sqrt{r}} \sum_{k=0}^{r-1} |u_k\rangle$

(a uniform superposition of the eigenvectors). Applying the eigenvalue estimation algorithm using this state will perform the map:

$$|0\rangle \left[\frac{1}{\sqrt{r}} \sum_{k=0}^{r-1} |u_k\rangle \right] \mapsto \frac{1}{\sqrt{r}} \sum_{k=0}^{r-1} |\widetilde{k/r}\rangle |u_k\rangle. \tag{3.21}$$

Here, we can measure the first register to get an estimate of k/r, where k is random integer between 0 and $r-1$.

We can use the state $|0\rangle|1\rangle$ instead of $|0\rangle \left[\frac{1}{\sqrt{r}} \sum_{k=0}^{r-1} |u_k\rangle \right]$ to do this without knowing r, since it is possible to show that $\frac{1}{\sqrt{r}} \sum_{k=0}^{r-1} |u_k\rangle = |1\rangle$. To demonstrate this, consider the amplitude of the state $|1\rangle$ in the superposition $\frac{1}{\sqrt{r}} \sum_{k=0}^{r-1} |u_k\rangle$. Let $a_{|m\rangle}$ denote the amplitude corresponding to $|m\rangle$ in this superposition. Since

$$\frac{1}{\sqrt{r}} \sum_{k=0}^{r-1} |u_k\rangle = \frac{1}{\sqrt{r}} \sum_{k=0}^{r-1} \frac{1}{\sqrt{r}} \sum_{s=0}^{r-1} e^{-2\pi i \frac{k}{r} s} |a^s \mod N\rangle \tag{3.22}$$

$$= \frac{1}{r} \sum_{k=0}^{r-1} \sum_{s=0}^{r-1} e^{-2\pi i \frac{k}{r} s} |a^s \mod N\rangle, \tag{3.23}$$

it is evident that

$$a_{|m\rangle} = \frac{1}{r} \sum_{k=0}^{r-1} \left[\sum_{\{s|a^s \mod N=m,\ s \le r-1\}} e^{-2\pi i \frac{k}{r} s} \right]. \tag{3.24}$$

To compute the amplitude of $|1\rangle$, note that $s \equiv 0 \mod r \iff |a^s \mod N\rangle = |1\rangle$. With this, we have

$$a_{|1\rangle} = \frac{1}{r} \sum_{k=0}^{r-1} \left[\sum_{\{s=0\}} e^{-2\pi i \frac{k}{r} s} \right] = \frac{1}{r} \sum_{k=0}^{r-1} e^{-2\pi i \frac{k}{r} 0} = \frac{1}{r} r = 1. \tag{3.25}$$

$$\tag{3.26}$$

Since $\sum_{m=0}^{r-1} a_{|m\rangle} = 1$, we can conclude that $a_{|m\rangle} = 0$ for all $m \ne 1$ which implies that $\frac{1}{\sqrt{r}} \sum_{k=0}^{r-1} |u_k\rangle = |1\rangle$ as needed.

3.2.2   ORDER FINDING: CONTINUED FRACTIONS

We just saw how measurement at the end of the eigenvalue estimation algorithm gave us an estimate of k/r. Later, we will see how this estimate is actually a number x such that $x/2^n$ is an estimate of k/r. Now we discuss how we can use this estimate to actually compute the order of the function in question.

To begin, let us review the concept of a continued fraction. A continued fraction $[a_0, ..., a_M]$ has the form

$$[a_0, ..., a_M] = a_0 + \cfrac{1}{a_1 + \cfrac{1}{a_2 + \cfrac{1}{... + \cfrac{1}{a_{a_M}}}}}. \tag{3.27}$$

where a_k is an integer for any k. The n-th convergent of a continued fraction p_n/q_n is defined as

$$[a_0, ..., a_n] = \frac{p_n}{q_n}. \tag{3.28}$$

It is possible to represent any rational number in terms of integers using continued fractions. In order to construct such a representation, we can use the continued fractions algorithm. This algorithm is stated below:

1. Start with an rational number a/b.

2. **Split** a/b is first split into its integer and fractional parts. Let c be the result of the a/b using integer division. Hence, we have:

$$\frac{a}{b} = c + \frac{a \mod b}{b}. \tag{3.29}$$

3. **Invert** The fractional component is then inverted. This gives:

$$\frac{a}{b} = c + \frac{1}{\frac{b}{a \mod b}}. \tag{3.30}$$

4. The procedure is then repeated with the fraction $\frac{b}{a \mod b}$ as the number a/b in the next step. Doing this repeatedly produces the continued fraction representation of a/b in $O(L)$ cycles where a and b are L-bit integers.

Why are we interested in the process of computing continued fraction representations and the convergents created in the process? The reason lies with the following theorem [149, 184]: if a fraction k/r satisfies

$$\left| \frac{x}{2^n} - \frac{k}{r} \right| \leq \frac{1}{2r^2} \tag{3.31}$$

then $\frac{k}{r}$ appears in the list of convergents of $\frac{x}{2^n}$. This will allow us to finally compute the period r. This can be easily done with two runs of the eigenvalue estimation algorithm. On the first run of the the eigenvalue estimation algorithm, we will measure some value $\frac{x_1}{2^n}$ and then use the continued fractions algorithm attain a value

$$\frac{k_1}{r} = \frac{c_1}{r_1} \text{ where } GCD(c_1, r_1) = 1. \tag{3.32}$$

Note that k_1 and r might share a common factor, so we may not be able to determine r by examining the resultant fraction. This motivates the need for a second run of the algorithm, where we measure another value $\frac{x_2}{2^n}$ which allows us to compute another fraction

$$\frac{k_2}{r} = \frac{c_2}{r_2} \text{ where } GCD(c_2, r_2) = 1. \tag{3.33}$$

Since both r_1 and r_2 divide r, we can now compute $r = LCM(r_1, r_2)$.

3.3   QUANTUM PHASE ESTIMATION

The goal of the phase estimation algorithm is to produce a good estimate of the phase parameter ω given the quantum state $|\psi\rangle = \frac{1}{2^n} \sum_{y=0}^{2^n-1} e^{2\pi i \omega y} |y\rangle$. Before we discuss the quantum phase estimation algorithm, let us first review two quantum operators: the quantum Fourier transform (QFT) and the inverse quantum Fourier transform (iQFT):

$$QFT \quad : \quad |x\rangle \mapsto \frac{1}{\sqrt{2^n}} \sum_{y=0}^{2^n-1} e^{2\pi i \frac{x}{2^n} y} |y\rangle \tag{3.34}$$

$$iQFT \quad : \quad |x\rangle \mapsto \frac{1}{\sqrt{2^n}} \sum_{y=0}^{2^n-1} -e^{2\pi i \frac{x}{2^n} y} |y\rangle \tag{3.35}$$

The quantum phase estimation algorithm simply consists of an application of the inverse QFT on the input state. Measuring the quantum state after this operation gives us an approximation of the phase term. The quantum circuit in Figure 3.2 implements the inverse QFT. We will provide overview of the proof of why this works in this section.

3.3.1   PROOF SKETCH OF THE CORRECTNESS OF THE PHASE ESTIMATION CIRCUIT

There are two kinds of gates in Figure 3.2: Hadamard gates and controlled inverse rotation gates. The single qubit Hadamard gate defined by the following matrix:

$$H = \frac{1}{\sqrt{2}} \begin{bmatrix} 1 & 1 \\ 1 & -1 \end{bmatrix} \tag{3.36}$$

A single qubit rotation gate R_k and its inverse R_k^{-1} are defined by:

$$R_k = \begin{bmatrix} 1 & 0 \\ 0 & e^{\frac{2\pi i}{2^k}} \end{bmatrix}, \quad R_k^{-1} = \begin{bmatrix} 1 & 0 \\ 0 & e^{\frac{-2\pi i}{2^k}} \end{bmatrix} \tag{3.37}$$

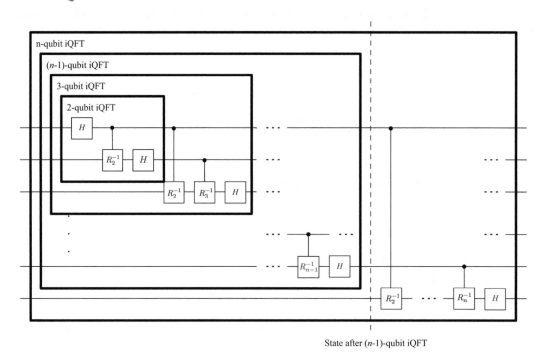

State after $(n\text{-}1)$-qubit iQFT

Figure 3.2: Inverse Quantum Fourier Transform (iQFT) circuit.

Given a n-qubit gate G that operates on an n-qubit state $|\psi\rangle$, the $n+1$-qubit controlled gate controlled $-G$ is defined as:

$$\text{controlled-}G : \begin{cases} |0\rangle|\psi\rangle & \mapsto & |0\rangle|\psi\rangle \\ |1\rangle|\psi\rangle & \mapsto & |1\rangle G|\psi\rangle \end{cases} \tag{3.38}$$

where G is a 1-qubit gate. With these definitions, we can now sketch of the proof that the circuit in Figure (3.2) implements the phase estimation algorithm. In this sketch, we will use the binary representation of the phase term $\omega = 0.x_1x_2...$ where $x_1, x_2, ...$ are binary values. We will segment this proof into two key parts (both of which can be done via induction) for clarity. These two pieces are described below.

1. **State decomposition lemma** At the beginning of the phase estimation algorithm, we start an n-qubit quantum state of the form $|\psi\rangle = \frac{1}{2^n} \sum_{y=0}^{2^n-1} e^{2\pi i \omega y}|y\rangle$. It is useful to attain the expression for the state of each of the individual qubits in order to analyze the circuit in Figure 3.2. The following relation gives us a way to express the state $|\psi\rangle$ as the tensor product

of n 1-qubit states:

$$\frac{1}{\sqrt{2^n}} \sum_{y=0}^{2^n-1} e^{2\pi i \omega y} |y\rangle \tag{3.39}$$

$$= \left(\frac{|0\rangle + e^{2\pi i (2^{n-1}\omega)} |1\rangle}{\sqrt{2}} \right) \tag{3.40}$$

$$\otimes \left(\frac{|0\rangle + e^{2\pi i (2^{n-2}\omega)} |1\rangle}{\sqrt{2}} \right) \otimes \dots \otimes \left(\frac{|0\rangle + e^{2\pi i (\omega)} |1\rangle}{\sqrt{2}} \right). \tag{3.41}$$

This relation is easily shown via induction. To show the base case (for $n = 1$), we simply expand the tensor representation on the left-hand side of the equation and re-form the sum on the right-hand side. For the induction step, we start with the relation for n-qubits and tensor both sides of the equation with the state of the $n + 1$-th qubit. Simplification of the resulting sum completes this step.

2. **Recursive iQFT circuit formulation** The circuit in Figure 3.2 can be defined recursively, as illustrated by the shaded areas of the diagram. Using this and the lemma from step 1, we can inductively demonstrate the operation of the circuit. We start by considering the base case of $n = 2$ qubits. This consists of a Hadamard operation on the first qubit, a controlled rotation operation on the other qubit (with the first qubit as the "control" line), and finally a second Hadmard operation on the second qubit. At the beginning of this circuit, the state of the system is:

$$\text{Initial state (2 qubits)} = \left(\frac{|0\rangle + e^{2\pi i (0.x_2)} |1\rangle}{\sqrt{2}} \right) \otimes \left(\frac{|0\rangle + e^{2\pi i (0.x_1 x_2)} |1\rangle}{\sqrt{2}} \right) \tag{3.42}$$

We can now show that applying the aforementioned sequence of operations on this initial state has the desired effect, as shown below:

$$\text{Initial state} \xrightarrow{\textit{H applied to qubit 1}} |x_2\rangle \left(\frac{|0\rangle + e^{2\pi i (0.x_1 x_2)} |1\rangle}{\sqrt{2}} \right) \tag{3.43}$$

$$\xrightarrow{\textit{Controlled rotation on qubit 2}} |x_2\rangle \left(\frac{|0\rangle + e^{2\pi i (0.x_1)} |1\rangle}{\sqrt{2}} \right) \tag{3.44}$$

$$\xrightarrow{\textit{H applied to qubit 2}} |x_2\rangle |x_1\rangle. \tag{3.45}$$

All that remains now is the induction step; we assume that the circuit works for n-qubits and proceed to show its operation for $n + 1$ qubits. To make the following manipulations easier, let's label the phase parameter (in binary), estimated by the n-qubit iQFT, is $\omega_n = 0.x_2 \dots x_{n+1}$. Our goal is to show that the $n + 1$-qubit iQFT will allow us to estimate the phase parameter

$\omega_{n+1} = 0.x_1 \ldots x_{n+1}$. Consider the state of the system after the n-qubit QFT (labeled as "State after $(n-1)$-qubit iQFT") in Figure 3.2:

$$|x_2\rangle \ldots |x_{n+1}\rangle \left(\frac{|0\rangle + e^{2\pi i(0.x_1\ldots x_{n+1})}|1\rangle}{\sqrt{2}} \right). \tag{3.46}$$

Applying the rest of the operations in the circuit (a sequence of controlled rotation gates and a Hadamard gate), we can compute the final state:

$$\text{State after } (n-1)-\text{qubit iQFT} \tag{3.47}$$

$$\xrightarrow{R_n^{-1}\ldots R_2^{-1} \text{ on qubit } n+1} |x_{n+1}\rangle \ldots |x_2\rangle \left(\frac{|0\rangle + e^{2\pi i(0.x_1)}|1\rangle}{\sqrt{2}} \right) \tag{3.48}$$

$$\xrightarrow{H \text{ applied to qubit } n+1} |x_{n+1}\rangle \ldots |x_2\rangle |x_1\rangle. \tag{3.49}$$

This is the phase state that we required and therefore the proof is complete.

If we measure the quantum state at the end of the phase estimation circuit shown in this section, we will measure the state encoding the phase parameter ω with probability 1 if ω is of the form $0.x_1 \ldots x_n$. The circuit is still useful for estimating other values of ω, but it does so less reliably. It can be shown that if $\frac{x}{2^n} \le \omega \le \frac{x+1}{2^n}$, then we will measure the states corresponding to the binary representations of x or $x+1$ with probability $\frac{8}{\pi^2}$. While this measurement probability is lower than 1, it is still good enough to be useful for a variety of applications, including period-finding.

3.4   EIGENVALUE ESTIMATION

Now that we've seen an efficient algorithm for phase estimation, we can go into the design of an algorithm for eigenvalue estimation. The eigenvalue estimation problem is as follows: estimate the parameter ω, given an quantum operator U with eigenstate $|\psi\rangle$ with eigenvalue $e^{2\pi i\omega}$. The solution to this problem is to use a sequence of controlled operators to build a quantum state, which we can then apply the phase estimation algorithm to determine ω.

To begin, consider a controlled variant of the operator U. Call this operator $c - U$. Let's look at what happens when we apply this operator to the target state $|\psi\rangle$ and a control qubit in the state $\frac{|0\rangle + |1\rangle}{\sqrt{2}}$:

$$c\text{-}U\left[\left(\frac{|0\rangle + |1\rangle}{\sqrt{2}}\right)|\psi\rangle\right] = c\text{-}U\left(\frac{|0\rangle|\psi\rangle}{\sqrt{2}}\right) + c\text{-}U\left(\frac{|1\rangle|\psi\rangle}{\sqrt{2}}\right) \tag{3.50}$$

$$= \frac{|0\rangle|\psi\rangle}{\sqrt{2}} + \frac{e^{2\pi i\omega}|1\rangle|\psi\rangle}{\sqrt{2}} \tag{3.51}$$

$$= \left(\frac{|0\rangle + e^{2\pi i\omega}|1\rangle}{\sqrt{2}}\right)|\psi\rangle \tag{3.52}$$

As we can see above, we have introduced the eigenvalue $e^{2\pi i\omega}$ into the state of the control register. If we had chosen an operator U^{2^j} instead of U, the state of the control register would have become

$$\frac{|0\rangle + e^{2\pi i(2^j\omega)}|1\rangle}{\sqrt{2}} \qquad (3.53)$$

since U^{2^j} has an eigenvalue $(e^{2\pi i\omega})^{2^j} = e^{2\pi i(2^j\omega)}$. The state of this control register should look eerily similar to the states of the individual qubits at the beginning of the phase estimation algorithm. In fact, if we set $j = n - m$, the resulting state after the controlled operations is precisely the state of the m-th qubit at the start of the phase estimation algorithm. If we prepare n control registers in the state $\frac{|0\rangle+|1\rangle}{\sqrt{2}}$ and apply U^{n-m} to the m-th qubit, we will have prepared an n-qubit state that we can supply as input to the phase estimation algorithm to get an estimate of the parameter ω. Figure 3.3 shows a circuit for doing this. Note that we can prepare the n control registers into the state

$$\left(\frac{|0\rangle + |1\rangle}{\sqrt{2}}\right)^{\otimes n} \qquad (3.54)$$

by applying the QFT to the state $|0\rangle^{\otimes n}$.

Note that in the order-finding algorithm described earlier, we can use this kind of circuit with $U = U_a = |sa \mod N\rangle$ to determine the eigenvalues of U_a. In that application, the sequence of controlled applications of $U_a^{2^j}$ is referred to as modular exponentiation. In many publications, modular exponentiation is not described as a sequence of controlled functions, but as the mapping

$$|z\rangle|y\rangle \mapsto |z\rangle|ya^z \mod N\rangle. \qquad (3.55)$$

How can we arrive at this mapping from the sequence of controlled operations we just saw? We can do this by analyzing what happens when we apply the controlled sequence of operations to the state vector $|y\rangle|z\rangle$ (as described in [149]):

$$|z\rangle|y\rangle \xrightarrow{\text{Controlled operations}} |z\rangle U^{z_n 2^{n-1}} \dots U^{z_1 2^0} |y\rangle \qquad (3.56)$$
$$= |z\rangle U^{z_n 2^{n-1}} \dots U^{z_2 2}|ya^{z_1 2^0} \mod N\rangle \qquad (3.57)$$
$$= |z\rangle U^{z_n 2^{n-1}} \dots U^{z_3 2^2}|ya^{z_2 2^1} a^{z_1 2^0} \mod N\rangle. \qquad (3.58)$$

Continuing in this way for all the controlled operations, we get

$$|z\rangle|y\rangle \xrightarrow{\text{Controlled operations}} |z\rangle|ya^{z_n 2^{n-1}} \dots a^{z_1 2^0} \mod N\rangle \qquad (3.59)$$
$$= |z\rangle|ya^z \mod N\rangle. \qquad (3.60)$$

The actual structure of a modular exponentiation circuit can vary depending on how we implement the operation with quantum adders and other components.

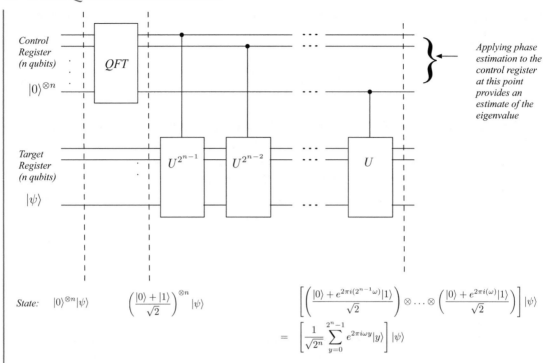

Figure 3.3: Phase state preparation circuit for eigenvalue estimation. Executing the phase estimation algorithm on the output of this circuit produces an estimate of the eigenvalue of the operator U.

3.5   THE HIDDEN SUBGROUP PROBLEM

As we mentioned earlier, the period-finding problem is an example of what is referred to in group theory as a Hidden Subgroup Problem (HSP). Efficient quantum algorithms are known for several other (but not all) hidden subgroup problems. In fact, some of the biggest open problems in the theory of quantum computation are concerned with whether or not efficient quantum algorithms exist for certain hidden subgroup problems.

A little bit of basic group theory is necessary to present the definition of the hidden subgroup problem. A set G is defined as a **group** over a binary **group operation** $\cdot$ if it satisfies the following properties:

- **Closure** The result of the group operation on any two elements is also in the group: for all $x, y \in G, x \cdot y \in G$.

- **Associativity** The group operation is associative: for all $x, y, z \in G, (x \cdot y) \cdot z = x \cdot (y \cdot z)$.

- **Identity** There exists an identity element $I \in G$ such that the result of the group operation with any element in the group and identity element produces the element in the group: for all $x \in G, x \cdot I = I \cdot x = x$.

- **Inverse** All elements $x \in G$ have an inverse $x^{-1} \in G$ such that the result of the group operation of any element with its inverse produces the identity: $x \cdot x^{-1} = x^{-1} \cdot x = I$

A group is **abelian** if for any $x, y \in G, x \cdot y = y \cdot x$. A **subgroup** H of a group G is a subset of the group that is also a group. We can use the $\leq$ symbol to signify that H is a subgroup of G: $H \leq G$. Given any subgroup $H \leq G$ and any element $x \in G$, the set $x \cdot H = \{x \cdot y : y \in H\}$ is a called a **left coset** of H and the set $H \cdot x = \{y \cdot x : y \in H\}$ is a called a **right coset** of H.

As an example, consider $\mathbb{Z}_8$, the additive group of the integers modulo 8. Table 3.1 shows the elements of this group in a table. The bolded elements in the table are elements of the subgroup $2\mathbb{Z}_8$ of $\mathbb{Z}_8$. These are the integers in this subgroup are the are multiples of 2 in the group. Both the group and subgroup in this example are abelian, as is illustrated by the diagonal symmetry of the group table.

Table 3.1: Example group and subgroup (subgroup elements are bolded)

Column element · Row element	0	1	2	3	4	5	6	7
0	**0**	1	**2**	3	**4**	5	**6**	7
1	1	2	3	4	5	**6**	7	0
2	**2**	3	**4**	5	**6**	7	**0**	1
3	3	4	5	**6**	7	**0**	1	2
4	**4**	5	**6**	7	**0**	1	**2**	3
5	5	**6**	7	**0**	1	**2**	3	4
6	**6**	7	**0**	1	**2**	3	**4**	5
7	7	**0**	1	**2**	3	**4**	5	**6**

The **hidden subgroup problem** (as defined in [110]) is as follows: Given a group G, a set X, and a map $f : G \to X$ such that there exists a "hidden subgroup" $S \leq G$ such that for all $x, y \in G$, $f(x) = f(y) \iff xS = yS$, determine the subgroup S. Figure 3.4 helps explain the properties of the subgroup S. All the cosets of S corresponding to the elements of G that map to the same element of X under f are constant and distinct from the cosets of S corresponding to the elements of G that map to different elements of X under f.

The problem of determining the order of the function $f(x) = a^x \mod N$ as defined earlier can be expressed as an HSP as follows:

- The group G is the group of integers modulo N under addition ($\mathbb{Z}_N$).

- The set X is also $\mathbb{Z}_N$.

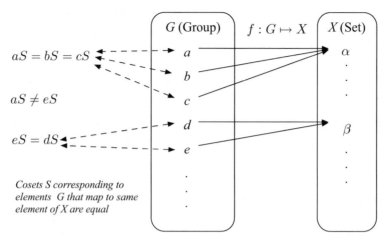

Figure 3.4: Properties of the subgroup S in the hidden subgroup problem.

- The map $f : G \rightarrow X$ is the function $f(x) = a^x \mod N$.

- The hidden subgroup S is $r\mathbb{Z}_N$, where r is the order of $f(x)$.

Let's consider the problem of determining the period of $f(x) = 3^x \mod 8$. G and X are then $\mathbb{Z}_8$ and $r = 2$ (which we can arrive at by evaluating f(x) for $0, ..., 7$) so $S = 2\mathbb{Z}_8$. Note that this group and subgroup are the same as in our basic group example. It is easy to see that $f(0) = f(2) = f(4) = f(6) = 1$, and $0 + 2\mathbb{Z}_8 = 2 + 2\mathbb{Z}_8 = 4 + 2\mathbb{Z}_8 = 6 + 2\mathbb{Z}_8 = \{0, 2, 4, 6\}$, and $f(1) = f(3) = f(5) = f(6) = 3$, and $1 + 2\mathbb{Z}_8 = 3 + 2\mathbb{Z}_8 = 5 + 2\mathbb{Z}_8 = 7 + 2\mathbb{Z}_8 = \{1, 3, 5, 7\}$.

Earlier, we saw an efficient quantum algorithm for solving this order-finding problem. It can be shown that any HSP on an Abelian group can be solved using an efficient quantum algorithm. The structure of these algorithms is expectedly similar to that of the period finding algorithm; a quantum Fourier transform (defined appropriately for the group in question), the application of a black-box function, an inverse quantum Fourier transform, and then classical steps to attempt to compute the generators of the hidden subgroup. Efficient quantum algorithms for solving arbitrary instances of the non-Abelian HSP are currently not known, although algorithms for some special cases have been found. A number of computational problems, including graph automorphism, can expressed as non-Abelian HSPs and therefore demonstrating the existence (or lack thereof) of such an algorithm would be particularly interesting from a quantum complexity theory perspective.

3.6   GROVER'S ALGORITHM FOR QUANTUM SEARCH

While it is common to hear of Grover's algorithm [89] as a quantum algorithm for "database search", the actual problem at hand is far more abstract than most actual classical database systems. This section describes the problem that Grover's algorithm solves, the specifics of the algorithm, and why it works.

3.6.1   SEARCHING WITH A QUANTUM BLACK BOX

Suppose that we are given a quantum black box U_f that implements some function $f(x) : \{0, 1\}^n \mapsto \{0, 1\}$. This is an $n + 1$-qubit operator operating on a $n + 1$ qubit state; the first n-qubits can be thought of as a target register while the last qubit can be thought of as a control register. $|x\rangle|y\rangle$. Its operation is shown below (where $|x\rangle$ is an n-qubit basis state and $|y\rangle$ is a 1-qubit basis state):

$$U_f : |x\rangle|y\rangle \mapsto |x\rangle|f(x) \oplus y\rangle. \tag{3.61}$$

As we can see, U_f maps an input state to an output state that encodes information about the values of the function $f(x)$. Observe that

$$f(x) = 0 \implies |f(x) \oplus 0\rangle = |0\rangle, |f(x) \oplus 1\rangle = |1\rangle \tag{3.62}$$
$$f(x) = 1 \implies |f(x) \oplus 0\rangle = |1\rangle, |f(x) \oplus 1\rangle = |0\rangle. \tag{3.63}$$

If we prepare the control register to the state $\frac{|0\rangle - |1\rangle}{\sqrt{2}}$ and apply U_f we can see the following effect:

$$|x\rangle \left(\frac{|0\rangle - |1\rangle}{\sqrt{2}} \right) \xrightarrow{U_f} \left(\frac{|x\rangle|f(x) \oplus 0\rangle - |x\rangle|f(x) \oplus 1\rangle}{\sqrt{2}} \right) \tag{3.64}$$

$$= \left(\frac{(-1)^{f(x)}|x\rangle|0\rangle - (-1)^{f(x)}|x\rangle|1\rangle}{\sqrt{2}} \right) \tag{3.65}$$

$$= (-1)^{f(x)}|x\rangle \left(\frac{|0\rangle - |1\rangle}{\sqrt{2}} \right) \tag{3.66}$$

If we discard the control register after applying U_f, we can induce the map

$$|x\rangle \xrightarrow{U_f} (-1)^{f(x)}|x\rangle. \tag{3.67}$$

In the Grover problem, we assume that the function $f(x)$ evaluates to 0 for all $x \in \{0, 1\}^n$, except for a single value. Our goal is to determine the distinct element $x \in \{0, 1\}^n$ such that $f(x) = 1$.

3.6.2   GROVER'S ALGORITHM

Classical algorithms (with classical black boxes) cannot solve this problem with less than a linear number of queries. Grover's algorithm, however, can achieve this feat with only $O(\sqrt{N})$ quantum queries (where $N = 2^n$ is the size of the domain of the function). The algorithm (which involves operations on an $n + 1$-qubit register consisting of an n-qubit target register and a 1-qubit control register) is described below.

1. **Initial State Preparation** Start with the state

$$|\psi\rangle = \frac{1}{\sqrt{2^n}} \sum_{x=0}^{2^n-1} |x\rangle \tag{3.68}$$

in the target register. As shown below, $|\psi\rangle$ is easily constructed from the state $|00\ldots0\rangle$ by using an n-qubit Hadmard gate, whose operation is identical to the QFT we saw when we looked at the phase estimation algorithm:

$$|\psi\rangle = H|00\ldots0\rangle = \frac{1}{\sqrt{2^n}}\sum_{x=0}^{2^n-1} e^{2\pi\frac{0}{2^n}x}|x\rangle = \frac{1}{\sqrt{2^n}}\sum_{x=0}^{2^n-1}|x\rangle. \tag{3.69}$$

2. **Grover Iteration** Apply the following routine $\lfloor\frac{\pi}{4}\sqrt{2^n}\rfloor$ times:

   (a) Prepare the control register to $\frac{|0\rangle-|1\rangle}{\sqrt{2}}$, apply the black-box function U_f, and then discard the control register to induce the map $|x\rangle \overset{U_f}{\longrightarrow} (-1)^{f(x)}|x\rangle$ on the target register as explained previously.

   (b) Apply an n-qubit Hadmard gate

   (c) Apply the operator $U_{0^\perp}$, which is defined as:

   $$U_{0^\perp} : \begin{cases} |x\rangle \mapsto -|x\rangle, & x \neq 0 \\ |0\rangle \mapsto |0\rangle \end{cases} \tag{3.70}$$

   (d) Apply the n-qubit Hadmard gate again

3. **Measurement** Finally, we measure the target register. The state that we measure should be (with high probability) the state corresponding to the element in the domain of $f(x)$ that produces a nonzero result.

   Figure 3.5 shows the quantum circuit corresponding to this algorithm.

3.6.3   PROOF SKETCH OF THE CORRECTNESS OF GROVER ITERATION

To demonstrate the repeated applications of the Grover iteration produces the state that (when measured) has a high probability of producing the desired result, we will trace the state of the quantum register through the different steps of the iteration.

The Grover iteration procedure simply consists of repeated applications of the operator $U_f(HU_{0^\perp}H)$ to the quantum state. First, define a number of states and subspaces that will make analyzing the application of $U_f(HU_{0^\perp}H)$ easier. Specifically, we will define two 2-dimensional subspaces of the n-qubit state space (whose dimension is 2^n). The first subspace will be used to understand the effect of U_f, while the second will be used to understand the effect of $HU_{0^\perp}H$. The following list describes the bases of these subspaces:

- **Understanding U_f:** Let w be the value of x such that $f(x) = 1$. $|w\rangle$ is therefore the quantum state that we want to measure at the end of the algorithm. Define the states $|\psi_{good}\rangle = |w\rangle$ and

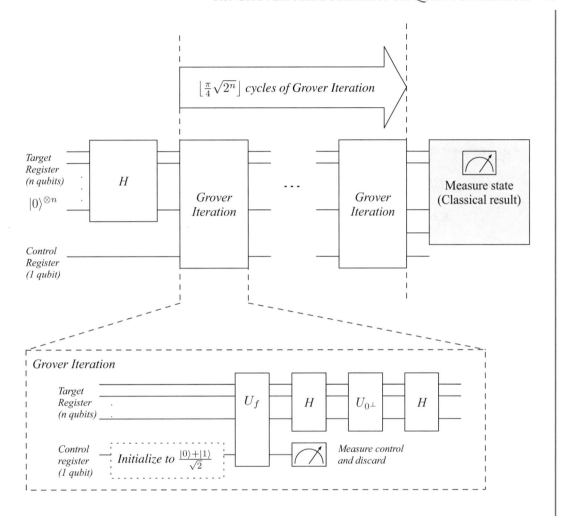

Figure 3.5: Quantum circuit for Grover's algorithm.

$|\psi_{bad}\rangle = \frac{1}{\sqrt{2^n-1}} \sum_{x \neq w} |x\rangle$ such that the state $|\psi\rangle$ at the end of the first step can be written as:

$$|\psi\rangle = \frac{1}{\sqrt{2^n}} |\psi_{good}\rangle + \sqrt{\frac{2^n - 1}{2^n}} |\psi_{bad}\rangle. \tag{3.71}$$

The effect of applying U_f to an element of the subspace spanned by $\{|\psi_{good}\rangle = |w\rangle, |\psi_{bad}\rangle\}$ is easy to see, as the mapping operation of U_f is defined in terms of $|\psi_{good}\rangle$ and $|\psi_{bad}\rangle$.

- **Understanding** $HU_{0\perp}H$**:** Let $|\bar{\psi}\rangle$ be the state that is orthogonal to $|\psi\rangle$. We can write $|\bar{\psi}\rangle$ as a linear combination of states of the form $H|x\rangle$ (where $x \neq 0$):

$$|\bar{\psi}\rangle = \sum_{k=1}^{2^n-1} a_k H|k\rangle, \ a_k \in \mathbb{R}. \tag{3.72}$$

Notice the result of the application of the operator $HU_{0\perp}H$ to $|\psi\rangle$ and $|\bar{\psi}\rangle$ (which occurs in the last three steps of Grover iteration):

$$HU_{0\perp}H|\psi\rangle \ = \ HU_{0\perp}H[H|00\ldots0\rangle] = H|00\ldots0\rangle = |\psi\rangle \tag{3.73}$$

$$HU_{0\perp}H|\bar{\psi}\rangle \ = \ HU_{0\perp}H\left[\sum_{k=1}^{2^n-1} a_k H|k\rangle\right] = \sum_{k=1}^{2^n-1} a_k HU_{0\perp}H[H|k\rangle] \tag{3.74}$$

$$= \ \sum_{k=1}^{2^n-1} a_k(-H|k\rangle) = -\sum_{k=1}^{2^n-1} a_k H|k\rangle = -|\bar{\psi}\rangle. \tag{3.75}$$

With these identities, we can easily quantify the effect of applying $HU_{0\perp}H$ to an element of the subspace spanned by $\{|\psi\rangle, |\bar{\psi}\rangle\}$.

Now let's analyze the Grover iteration process. Start with the state ψ at the beginning of the Grover iteration process, expressed in terms of $|\psi_{good}\rangle$ and $|\psi_{bad}\rangle$ (since we'll be applying U_f to this state first):

$$|\psi\rangle = \frac{1}{\sqrt{2^n}}|\psi_{good}\rangle + \sqrt{\frac{2^n-1}{2^n}}|\psi_{bad}\rangle. \tag{3.76}$$

The coefficients of each of the basis states can be expressed in terms of sines and cosines. Select an angle θ such that $\sin(\theta) = \frac{1}{\sqrt{2^n}}$. By the identity $\sin^2\theta + \cos^2\theta = 1$, this gives $\cos(\theta) = \sqrt{\frac{2^n-1}{2^n}}$, which conveniently is the coefficient of $|\psi_{bad}\rangle$ in the previous equation. With this, $|\psi\rangle$ becomes

$$|\psi\rangle = \sin(\theta)|\psi_{good}\rangle + \cos(\theta)|\psi_{bad}\rangle. \tag{3.77}$$

Applying U_f to this state, we get

$$U_f|\psi\rangle \ = \ U_f[\sin(\theta)|\psi_{good}\rangle + \cos(\theta)|\psi_{bad}\rangle] \tag{3.78}$$

$$= \ -\sin(\theta)|\psi_{good}\rangle + \cos(\theta)|\psi_{bad}\rangle. \tag{3.79}$$

We're now ready to apply $HU_{0\perp}H$ to this state. Before we do this, we will first express this state in terms of the states $\{|\psi\rangle$ and $|\bar{\psi}\rangle$. Table 3.2 shows a set of expressions for changing bases between the two sets we defined earlier.

Table 3.2: Basis identities for the proof sketch of the correctness of Grover's algorithm

State	Representation in other basis
$|\psi\rangle$	$\sin(\theta)|\psi_{good}\rangle + \cos(\theta)|\psi_{bad}\rangle$
$|\bar{\psi}\rangle$	$\cos(\theta)|\psi_{good}\rangle - \sin(\theta)|\psi_{bad}\rangle$
$|\psi_{good}\rangle$	$\sin(\theta)|\psi\rangle + \cos(\theta)|\bar{\psi}\rangle$
$|\psi_{bad}\rangle$	$\cos(\theta)|\psi\rangle - \sin(\theta)|\bar{\psi}\rangle$

Applying the basis identities in Table 3.2 and basic product-to-sum trigonometric identities, we attain

$$
\begin{aligned}
U_f|\psi\rangle &= -\sin(\theta)|\psi_{good}\rangle + \cos(\theta)|\psi_{bad}\rangle & (3.80)\\
&= -\sin(\theta)\sin(\theta)|\psi\rangle - \sin(\theta)\cos(\theta)|\bar{\psi}\rangle & (3.81)\\
&\quad + \cos(\theta)\cos(\theta)|\psi\rangle - \cos(\theta)\sin(\theta)|\bar{\psi}\rangle & (3.82)\\
&= \cos(2\theta)|\psi\rangle - \sin(2\theta)|\bar{\psi}\rangle. & (3.83)
\end{aligned}
$$

At this point, we can apply $HU_{0\perp}H$ to get

$$
\begin{aligned}
HU_{0\perp}HU_f|\psi\rangle &= HU_{0\perp}H\left[\cos(2\theta)|\psi\rangle - \sin(2\theta)|\bar{\psi}\rangle\right] & (3.84)\\
&= \cos(2\theta)|\psi\rangle + \sin(2\theta)|\bar{\psi}\rangle & (3.85)\\
&= \cos(2\theta)\left[\sin(\theta)|\psi_{good}\rangle + \cos(\theta)|\psi_{bad}\rangle\right] & (3.86)\\
&\quad + \sin(2\theta)\left[\cos(\theta)|\psi_{good}\rangle - \sin(\theta)|\psi_{bad}\rangle\right]\\
&= \left[\frac{\sin(2\theta + \theta) - \sin(2\theta - \theta)}{2}\right]|\psi_{good}\rangle & (3.87)\\
&\quad + \left[\frac{\cos(2\theta - \theta) + \cos(2\theta + \theta)}{2}\right]|\psi_{bad}\rangle\\
&\quad + \left[\frac{\sin(2\theta + \theta) + \sin(2\theta - \theta)}{2}\right]|\psi_{good}\rangle\\
&\quad - \left[\frac{\cos(2\theta - \theta) - \cos(2\theta + \theta)}{2}\right]|\psi_{bad}\rangle\\
&= \sin(3\theta)|\psi_{good}\rangle + \cos(3\theta)|\psi_{bad}\rangle. & (3.88)
\end{aligned}
$$

This gives us an expression for $HU_{0\perp}HU_f|\psi\rangle$ (the state after a single iteration of the Grover iteration procedure) in both basis states. But we are really interested in the state of the system after multiple Grover iterations. Luckily, it is easy to show that for any integer $k > 1$ the following relationship holds:

$$
\begin{aligned}
(HU_{0\perp}HU_f)^k|\psi\rangle &= \cos(2k\theta)|\psi\rangle + \sin(2k\theta)|\bar{\psi}\rangle & (3.89)\\
&= \sin((2k+1)\theta)|\psi_{good}\rangle + \cos((2k+1)\theta)|\psi_{bad}\rangle. & (3.90)
\end{aligned}
$$

We can use induction to do this. We've already done the base case (for $k = 1$) and so we need to focus on the induction step (proving the equality for $k + 1$ given that it is true for k). Doing this is very similar to the base case: We start with the previous equation and first apply U_f:

$$
\begin{aligned}
U_f(HU_{0\perp}HU_f)^k|\psi\rangle &= U_f\left[\sin((2k+1)\theta)|\psi_{good}\rangle + \cos((2k+1)\theta)|\psi_{bad}\rangle\right] \quad (3.91) \\
&= -\sin((2k+1)\theta)|\psi_{good}\rangle + \cos((2k+1)\theta)|\psi_{bad}\rangle. \quad (3.92)
\end{aligned}
$$

Then applying $HU_{0\perp}H$, we get

$$
\begin{aligned}
&(HU_{0\perp}H)[U_f(HU_{0\perp}HU_f)^k|\psi\rangle] \quad (3.93) \\
=\ &HU_{0\perp}H\left[-\sin((2k+1)\theta)|\psi_{good}\rangle + \cos((2k+1)\theta)|\psi_{bad}\rangle\right] \quad (3.94) \\
=\ &HU_{0\perp}H\left\{-\sin((2k+1)\theta)\left[\sin(\theta)|\psi\rangle + \cos(\theta)|\bar\psi\rangle\right]\right. \quad (3.95) \\
&\left. + \cos((2k+1)\theta)\left[\cos(\theta)|\psi\rangle - \sin(\theta)|\bar\psi\rangle\right]\right\} \\
=\ &-\sin((2k+1)\theta)\left[\sin(\theta)|\psi\rangle - \cos(\theta)|\bar\psi\rangle\right] \quad (3.96) \\
&+ \cos((2k+1)\theta)\left[\cos(\theta)|\psi\rangle + \sin(\theta)|\bar\psi\rangle\right] \\
=\ &-\left[\frac{\cos((2k+1)\theta - \theta) - \cos((2k+1)\theta + \theta)}{2}\right]|\psi\rangle \quad (3.97) \\
&-\left[\frac{\sin((2k+1)\theta + \theta) + \sin((2k+1)\theta - \theta)}{2}\right]|\bar\psi\rangle \\
&+\left[\frac{\cos((2k+1)\theta - \theta) + \cos((2k+1)\theta + \theta)}{2}\right]|\psi\rangle \\
&-\left[\frac{\sin((2k+1)\theta + \theta) - \sin((2k+1)\theta - \theta)}{2}\right]|\bar\psi\rangle \\
=\ &\cos(2(k+1)\theta)|\psi\rangle + \sin(2(k+1)\theta)|\bar\psi\rangle. \quad (3.98)
\end{aligned}
$$

Switching bases, this is easily shown to be equivalent to $\sin((2(k+1)+1)\theta)|\psi_{good}\rangle + \cos((2(k+1)+1)\theta)|\psi_{bad}\rangle$. Hence, we are done with the induction step and the proof is complete.

Figure 3.6 shows a visual representation of the relationship between the state of the quantum system and the basis states $\{|\psi_{good}\rangle, |\psi_{bad}\rangle\}$ through several cycles of the Grover iterations procedure. If we label the x and y axes of a plot of $\mathbb{R}^2$ by $|\psi_{bad}\rangle$ and $|\psi_{good}\rangle$, respectively, we can visualize the state of the system after k iterations $(\sin((2k+1)\theta)|\psi_{good}\rangle + \cos((2k+1)\theta)|\psi_{bad}\rangle)$ as a vector with phase $(2k+1)\theta$. Every stage of Grover iteration further "rotates" this vector. Since our goal is to measure the state $|\psi_{good}\rangle$, we want to rotate the state vector so that it is as close to $|\psi_{good}\rangle$. This occurs when $\sin((2k+1)\theta)$ is close to 1 and $\cos((2k+1)\theta)$ is close to 1. Solving for k and analyzing the probability that we get the correct answer, it can be shown [110] that it requires $k = \lfloor\frac{\pi}{4}\sqrt{2^n}\rfloor$ queries to measure the state $|\psi_{good}\rangle$ with probability $1 - O(\frac{1}{2^n})$.

3.7   QUANTUM ADIABATIC ALGORITHMS

One of the most important elements of the theory of quantum mechanics is the model of how quantum systems evolve. The behavior of a quantum state $|\psi(t)\rangle$ over time is described by the

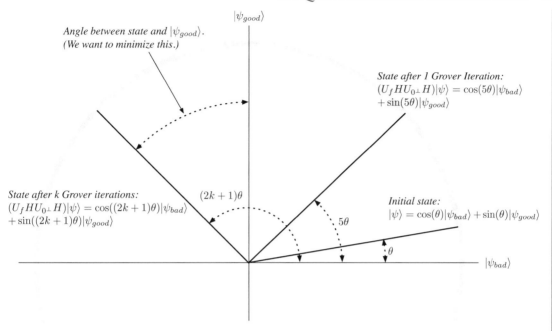

Figure 3.6: Visual representation of the effect of Grover Iterations on $|\psi\rangle$.

Schrodinger equation:

$$i\frac{d}{dt}|\psi(t)\rangle = H(t)|\psi(t)\rangle. \tag{3.99}$$

where $H(t)$ is a time dependent Hamiltonian [75]. In quantum adiabatic computation, we perform "computation" via the transition of the Hamiltonian from one state (H_{start}) to another (H_{end}) as governed by the Schrodinger equation. In particular, we are interested in the ground states (eigenvectors corresponding to the smallest eigenvalue) of each operator. The **quantum adiabatic theorem** states that if we evolve the system for a sufficiently long time, the ground state of H_{start} will be transformed to the ground state of H_{end}. Suppose we select H_{end} such that its ground state encodes the problem solution and H_{start} is such that its ground state is easy to find. Now if we evolve the system for the appropriate amount of time, we can determine the solution to the problem via the ground state of the final operator. Figure 3.7 shows a visual summary of the principle behind quantum adiabatic algorithms.

There are several results that attempt to precisely bound the time T required to evolve the system to solve problems in this manner. One such bound is stated as follows. Let $\tilde{H}(s)$ be the state of the of the Hamiltonian at time s, where $0 \leq s \leq 1$ (note that here $\tilde{H}(0) = H_{start}$ and $\tilde{H}(1) = H_{end}$). Here, we assume that the first and second derivatives exist and are bounded. Now

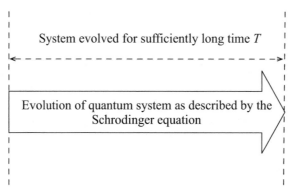

Starting Hamiltonian H_{start} (ground state is easy to construct)

Ending Hamiltonian H_{end} (ground state encodes solution)

Figure 3.7: High-level overview of adiabatic quantum computation.

choose

$$H(t) = \tilde{H}(\frac{t}{T}), t \in [0, 1].$$ (3.100)

Ambainis et al. [9] demonstrate that for any $\delta > 0$, if we select

$$T \geq \frac{10^5}{\delta^2} \max \left\{ \frac{||H'||^3}{\lambda^4}, \frac{||H||'||H||''}{\lambda^3} \right\}$$ (3.101)

where λ is the minimum eigenvalue gap of $\tilde{H}$, the final state is at most distance δ from the ground state of the end Hamiltonian.

3.7.1   3-SAT: AN EXAMPLE OF A QUANTUM ADIABATIC ALGORITHM

3-SAT is the problem of finding a valid set of assignments to N Boolean variables $x_1, \ldots, x_N$ that satisfy a set of M Boolean clauses. Each clause of C has the following form:

$$C = y_{C1} \vee y_{C2} \vee y_{C3},$$ (3.102)

where $y_{Ck} \in \{x_1, \ldots, x_n, \neg x_1, \ldots, \neg x_n\}$. Hence, we can also state 3-SAT as the problem of finding $x_1, \ldots, x_N$ such that the Boolean function $f(x_1, \ldots, x_n) = C_1 \wedge \ldots \wedge C_k$ evaluates to true.

3-SAT is an intriguing problem in computer science for numerous reasons. Key among these is that it belongs to the complexity class **NP-complete**. As of this writing, the existence of polynomial time algorithms for solving NP-complete on both classical and quantum computers is currently not known.

In 2000, Farhi et al. [75] proposed an adiabatic quantum algorithm for solving **unique** 3-SAT, where we are guaranteed a single unique solution to the 3-SAT instance. While computer simulations on random instances of 3-SAT by Farhi et al. [75, 76] initially suggested a polynomial relationship between the time T required to evolve the system and the number N of Boolean variables in the problem, Van Dam et al. later demonstrated that this is actually not the case for all instances of the problem [210]. However (as also shown by Van Dam et al.), this does not rule out the possibility of polynomial time quantum adiabatic algorithms for NP-complete algorithms.

To further demystify the concept of adiabatic quantum algorithms, we will look at the structure of Hamiltonians H_{start} and H_{end} in the adiabatic quantum algorithm proposed by Farhi et al.

- **Ending Hamiltonian** Let's begin by analyzing the ending Hamiltonian H_{end} in Farhi's algorithm. The ground state of this Hamiltonian should encode a solution to the 3-SAT problem. Let the 2^n possible assignments of the variables $x_1, \ldots, x_n$ correspond to basis states $|x_1 \ldots x_n\rangle$ of the system. Define to be the operator H_k corresponding to the clause C of the 3-SAT formula as

$$H_k|x_1 \ldots x_n\rangle = h_C|x_1 \ldots x_n\rangle, \tag{3.103}$$

where h_C is 1 if $(x_1 \ldots x_n)$ does not satisfy clause C and 0 otherwise. Let H_{end} be the sum of all the operators H_k

$$H_{end} = \sum_C H_C. \tag{3.104}$$

Consider the effect of this operator on a basis vector $|x_1 \ldots x_n\rangle$:

$$H_{end}|x_1 \ldots x_n\rangle = \sum_C H_C|x_1 \ldots x_n\rangle. \tag{3.105}$$

The value of the sum is the number of clauses that $|x_1 \ldots x_n\rangle$ violates. Call this integer $\lambda_{|x_1 \ldots x_n\rangle}$. So H_{end} is now

$$H_{end}|x_1 \ldots x_n\rangle = \lambda_{|x_1 \ldots x_n\rangle}|x_1 \ldots x_n\rangle. \tag{3.106}$$

From this, we see that every basis state $|x_1 \ldots x_n\rangle$ is an eigenvector of H_{end} with corresponding eigenvalue $\lambda_{|x_1 \ldots x_n\rangle}$. The ground state of this operator is the basis vector that corresponds to the assignment of Boolean variables that satisfies all the clauses and the ground state energy is 0. This is guaranteed to be the ground state since all the other eigenvectors correspond to assignments of the Boolean variables that violate one or more clause.

- **Starting Hamiltonian** Let $H_B^{(i)}$ be an operator that acts on the i-th bit of the basis state and let d_i denote the number of times bit i appears in the clauses of the 3-SAT instance in question. $H_B^{(i)}$ is defined as

$$H_B^{(i)} = I^{\otimes(i-1)} \otimes \frac{1}{2}\begin{bmatrix} 1 & -1 \\ -1 & 1 \end{bmatrix} \otimes I^{\otimes(N-i)}. \tag{3.107}$$

With this, we can define H_{start} as

$$H_{start} = \sum_{i=1}^{N} d_i H_B^{(i)}.$$

(3.108)

Using the eigenvalues and eigenvectors of

$$\frac{1}{2} \begin{bmatrix} 1 & -1 \\ -1 & 1 \end{bmatrix}$$

(3.109)

and the eigenvalues and eigenvectors of the identity operators, we can deduce that the only eigenvector of H_{start} with eigenvalue 0 is

$$\left[\frac{1}{\sqrt{2}} |0\rangle + \frac{1}{\sqrt{2}} |1\rangle \right]^{\otimes N} = \frac{1}{\sqrt{2^n}} \sum_{x_1=0}^{1} \sum_{x_2=0}^{1} \cdots \sum_{x_n=0}^{1} |x_1 \ldots x_n\rangle.$$

(3.110)

and all the other eigenvectors have positive eigenvalues. Hence, this is the ground state of the starting Hamiltonian of the system.

CHAPTER 4

Building Reliable and Scalable Quantum Architectures

Designing faster and physically smaller computer architectures is desirable in both the quantum and classical computing domains. The impact of these two design parameters, however, has special significance in the quantum world. In classical computing, slower clock speed and larger gate size usually implies a more reliable chip that is less susceptible to errors. On the other hand, faster clock speeds, in quantum computing, limit the time that each qubit decoheres, which could make the architecture more reliable. Similarly, minimizing the quantum computer chip area reduces the physical distances that the quantum information must travel, which again could make the underlying qubit registers in the architecture less susceptible to errors.

A high-level schematic of an example quantum computer architecture implementing the Quantum Random Access Machine [118] model for quantum computing is shown in Figure 4.1. The QRAM model is an extension of the classical Random Access Machine model and consists of a classical subsystem responsible for generating the quantum operations that are executed on a quantum subsystem. The architecture in Figure 4.1 represents the quantum subsystem and is composed of a number of processing elements (PE) where each PE is designed to execute a *localized* piece of the larger application. The interconnect facilitates reliable PE-to-PE communication by allowing individual qubits to be either physically transported between different PE's or to be teleported between different PE's. If the PE's are sufficiently small (in terms of area), communication between qubits within each PE can be accomplished by direct physical qubit movement (as allowed by the underlying technology). Classical control processors orchestrate the scheduling of quantum operations, and the only means of communication between the classical and quantum hardware is through measurement results. Quantum error correction codes may be employed within each PE and within the design of the interconnect to limit the effects of decoherence on the individual qubits.

Following the successful implementation of physical qubits and quantum gates, DiVincenzo [67] published a set of hardware requirements that if met by a given device technology, then it could, in principle, be used to build a working computer. The set of criteria proposed by DiVincenzo can be summarized with the following four bullet points:

- A quantum register, described as a collection of well defined single-qubit states, must be initialized to a well known starting state, such as the state $|00 \dots 0\rangle$.

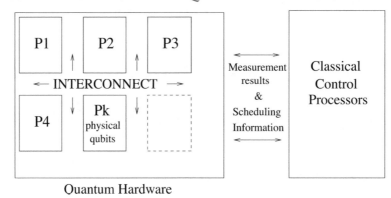

Figure 4.1: High Level Schematic for a Quantum Computer Architecture, based on the Quantum Random Access Machine model.

- A "universal" set of quantum logic gates must be available, where the gate cycle time must be much shorter than the relevant decoherence time cycle of the quantum register.

- Reliable measurements must be performed on any single-qubit state.

- The ability to transmit quantum information between specified locations, either through the direct physical movement of the qubit or by passing the information to "flying" qubits, which can then pass it back to "stationary" qubits for computation.

There is a difference, however, between being able to physically implement the necessary components needed for a quantum computer and designing a complete, large-scale quantum architecture that is intended to execute computationally relevant programs. Previous work in large-scale quantum architecture design, based on the circuit model [57, 78, 140, 153, 154], has allowed us to identify three main areas where significant challenges exist for building scalable architectures:

1. **Reliable and realistic implementation technology** that adheres to the DiVincenzo requirements [67] for implementing quantum computation.

2. **Robust, fault-tolerant structures** encoded using efficient error correction algorithms. This requirement provides system-level fault-tolerance that will allow the execution of an arbitrarily large sequence of universal quantum logic operations within the architecture decoherence time.

3. **Efficient quantum resource distribution** at both the application level and the physical qubit level that allows maximum overlap of computation and error correction.

Each of the three challenge areas is reviewed in detail in the subsequent subsections of this chapter. In Section 4.1, the existing technologies for realizing large-scale quantum computation are examined, and an overview of some of the components for trapped ion quantum computation and optical quantum computers is given. In Section 4.2, we show how quantum error correction codes are used to build scalable, fault-tolerant architectures. In Section 4.3, we provide an overview of the methods for quantum data distribution.

4.1   RELIABLE AND REALISTIC IMPLEMENTATION TECHNOLOGY

The physical realization of the basic components necessary for universal quantum computation has gathered much attention in recent years, and realistic technologies have emerged that have made the concept of quantum computing a feasible prospect. There are now physical schemes using trapped atomic ions [18, 169, 171] that have demonstrated every major low-level architectural component needed for scalable computing. Designs for large-scale systems employing quantum dots [204], polar molecules [166], and superconductors [28] are on the drawing board. Most recently, the CNOT gate has been experimentally demonstrated with superconducting qubits [96, 151], and a simple implementation of Shor's quantum factoring algorithm has been demonstrated using photonic quantum circuits [135]. Even so, the engineering difficulties associated with existing QIP technologies are significant, and the technology models vary to such an extent that identifying a clear winner at such an early stage may adversely affect future progress in building quantum computers.

The major limitations for technologies that can be used to build quantum hardware for a large-scale computer are two-fold:

- A number of qubits must be prepared and isolated from the environment such that they are protected from external forces that cause decoherence. Since completely isolating qubits is impossible, each physical qubit has a limited coherence time before its quantum state is lost. The coherence time is defined as the time during which a quantum state holds its superposition.

- The second major difficulty for emerging quantum technologies is the fundamental inability to copy a quantum state, combined with the need to perform logic operations and measurement on any arbitrary qubits.

The second limitation is particularly difficult to overcome. Classical data can be replicated using a FANOUT gate and transmitted on wires from the memory elements to the processing units. An imperfect classical gate or a leaking wire may have some effect on parts of the classical state, but usually not enough to outweigh the multitude of electrons used to encode a single bit of information. On the other hand, to operate on a number of qubits, we must be able to build them so that they are not only extremely weakly coupled to external decoherence forces, but also strongly coupled to each other and to external gate devices for the duration of a quantum logic gate. In addition, in order to transmit quantum information without leaving any trace behind, the information must be carefully protected while physically moving.

Physical implementations of qubits that move the quantum information easily and ones that allow state manipulation easily are two very contradicting concepts. For example, a qubit defined by the polarization states of photons is ideal for movement because it does not interact with its environment easily and it moves very fast. Photons, however, are hard to contain and two-qubit gates are very difficult to implement since photons do not interact with anything very well (including each other). On the other hand, heavy atoms are ideal for computation because they are relatively easy to slow down and apply operations on (usually through lasers), but they are difficult to transport. A middle ground qubit is one that is not only exposed to the environment for computation, but also moves with relative ease and speed. Unfortunately, the qubit's availability for computation also exposes it to uncontrollable forces during both computation and movement, making any choice for a qubit a choice with fundamentally limited reliability and decoherence time.

Physical realizations of the circuit model of quantum computation divide into several experimental proposals from very diverse fields of physical science, such as nuclear magnetic resonance (NMR) quantum computation [58, 82]; ion trap quantum computation [53] both optically through the coupling of neutral atoms with photons [29, 42] and physical segmented traps [38, 112, 221]; cavity quantum electro-dynamic (QED) computation [206]; optical quantum computation [123, 143]; solid state spin-based quantum computation [108, 109, 129, 188]; quantum dots [32, 136]; and "unique" qubits such as electrons floating on liquid helium [161], the quantum Hall effect [165], and qubits encoded in the charge distribution of a single electron on two donors [98]. The distinguishing feature for all proposed technologies is the implementation of the qubit, which, in turn, guides the control infrastructure of the computer itself and the speed of the computer.

Each approach has different strengths and weaknesses for implementing a truly scalable computer. For example, the Kane technology approach [8, 188], where qubits are realized using the electronic states of phosphorus atoms embedded in a silicon substrate, has the advantage that it builds upon existing investments in silicon fabrication techniques. Current measurement methods, however, can take as long as 4 days (far longer than the qubit coherence time of less than 60 ms), and there are also no existing experimental implementation of logic gates.

In another well developed work, qubits are held in pairs of energy levels of ions trapped in space by the electric potentials of metal electrodes [53, 112]. The ion-trap scheme is the only technology where every universal element for quantum computation has been realized with a clear scalable communication model [18, 171]. The caveat is that the ion-trap scheme is spatially expensive in terms of area and it requires the orchestration of millions of ions separately addressed via laser pulses, so it is not clear if it will remain a good technology for universal quantum computation in the distant future. For this reason, it is important to realize that the relevance of a certain technology must be judged as much for its potential promise as for its current experimental state. Reference [215] offers a very comprehensive review of the available technologies and their current parameters that are useful for building small computer prototypes. Here we give a brief description of two of the most successful experimental techniques so far: optical quantum computers and trapped-ion quantum computers.

4.1.1   OPTICAL QUANTUM COMPUTATION: PHOTONS AS QUBITS

The importance of photons as qubits is evident in their application to experimentally and commercially realize quantum cryptography protocols [24, 83]. In addition, photon-based qubits are the first physical system used to experimentally demonstrate entanglement [126, 128, 155], teleportation [35, 36, 80, 113, 157], various small-scale quantum algorithms [99, 127, 203], and even small instances of quantum factoring [162]. Photonic qubits are also the underlying device technologies in existing quantum architecture designs based on the cluster-state model of quantum computation [78, 148, 227].

The photon is the smallest unit for quantum information and has the advantage that it is virtually free of decoherence when implementing single-qubit gates. This stability is also the source of a severe experimental challenge since quantum information tends to be "trapped" in the photon making two-qubit gates very difficult to utilize with sufficient success rate. This means that photons do not interact easily with each other, making them difficult to realize for scalable quantum computation but ideal for quantum key distribution. An excellent review on optical quantum computation can be found as Reference [222].

Knill, Laflamme, and Milburn developed a scheme in 2001 [123] which demonstrates that, in principle, it is possible to create highly efficient scalable quantum computers using linear optical components made up of phase shifters and beam splitters, single photons, and photo detectors with only a polynomial resource overhead. Before this scheme was proposed, it had been shown that any unitary operator can be realized with linear optical components [167]. Single qubit operators are relatively straight forward with beam splitters and phase shifters, which are mathematically described as 2×2 operators along the Z and Y axis of the qubit state representation. As shown in Reference [149], any single unitary operator can be decomposed to a combination of Z and Y rotations.

For two-qubit gates, linear optics alone is not sufficient. Photo detectors are used to perform measurement, which combined with teleportation can be utilized to implement two qubit operations. The most reliable CNOT implementation with linear optics succeeds with probability of nearly 7% [119], while single-qubit operations are virtually noiseless. The use of a generalized beam splitter, combined with the Fourier transform, can help bring the success of the teleportation procedure close to unity at the expense of exponential photon resources. Scalable quantum computation is possible when the reliability of two-qubit gates through teleportation is reduced to a level such that quantum error correction can be utilized [119, 120].

Photon-based qubits offer ideal environment for distributed quantum computation. Photons are an attractive medium for shuttling quantum information from one part of the processor to another where the processor itself is composed of qubits that allow efficient quantum computation such as ion-traps. Optical qubits offer, by far, the most advance experimental implementation for entangling two remote ions [31, 71, 145] or inducing qubit-qubit interactions between solid-state qubits, using a common laser beam acting as a shared quantum bus [138, 192]. Entangling two remote qubits will allow the transfer of quantum information from one location to another through the teleportation procedure. Recently, linear optics quantum computation received a significant boost with the development of cluster-state quantum computation [37, 148, 150, 227], where initially entangled states are created that represent every necessary qubit resource in the architecture. Through single-qubit measurements on the entangled states, arbitrary quantum circuits can be simulated.

4.1.2   TRAPPED-ION QUANTUM COMPUTATION: IONS AS QUBITS

Recent experiments with trapped atomic ions have demonstrated all necessary quantum hardware components for building large-scale quantum computers. The idea for using trapped ions was initially proposed by Cirac and Zoller in 1995 [53] where a number of atomic ions are acted upon by lasers to implement quantum logic. Quantum data is stored in the internal nuclear and electronic states of the ions, while the traps themselves are segmented metal traps (or electrodes) that allow individual ions to be addressed. The electrodes are placed typically on a 2D alumina substrate, together with the needed electronics that control the trapping potentials. Ions in neighboring traps can couple to each other, forming a linear chain of ions whose vibrational modes provide the qubit-qubit interactions needed by multi-qubit quantum gates [132, 191]. Together with single bit rotations, this yields a universal set of quantum logic. All quantum logic operations are implemented by applying lasers to the target ions, including measurement of the quantum state [90, 112, 163, 221]. Multiple ions in different trap arrays can be controlled in parallel by focusing lasers through MEMS mirror arrays [114]. Additional *sympathetic cooling* ions are used to absorb unwanted vibrations from data ions, which are then dampened through laser manipulation [29, 30].

Figure 4.2 shows a schematic of the physical structure of a trap element in an ion-trap computer. In Figure 4.2(a), we see a single ion-group trapped in the middle trapping region. An ion-group will be abstracted as an inseparable pair of a data ion and a sympathetic cooling ion that will always move together. In reality, it may be technologically infeasible to implement reliable two-qubit quantum operations with the cooling ions present between the data ions; in which case the cooling ions must be provided separately. Trapping regions are the locations where ions can be prepared for the execution of a logical gate operation, which is simply an external laser source shining on the ion-group. Figure 4.2(b) demonstrates an ion-group moving from the top left trapping region to the middle region where a two-bit logical operation is to be performed. A fundamental time-step, or a clock cycle, in an ion-trap computer will be defined as any logical operation (one-bit or two-bit), a basic move operation from one trapping region to another, and measurement. It has been suggested in the literature that optimistic expectations for the probability of failure for fundamental operations in ion-traps is in the order of 10^{-7} [156], and average physical cycle time of approximately $10 \, \mu s$ [193, 222]. This cycle time has been shown to be sufficient for cooling and the additional join and split operations needed for each fundamental operation [193].

Recent experiments that realize quantum teleportation using trapped ions [18, 171] have demonstrated all the necessary elementary components needed to build a large-scale ion-trap processor such as ions trapped in segmented electrode structures, laser induced ion cooling and manipulation, measurement using a pump laser that causes a state dependent scattering of photons from the ion, and, finally, the ability to move ions around by changing the trapping potentials. In

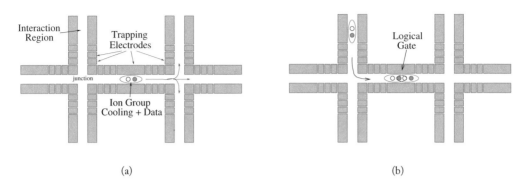

Figure 4.2: (a) The physical structure of an ion-trap quantum computer. An optimistic size of the trapping electrodes is in the order of tens of micrometers [193]. The data ion is kept together with a cooling ion and cooled before and after each movement step or logic gate. The ion-group can move to any of the 6 adjacent trapping regions for interaction with another ion-group.(b) A two-qubit gate sequence where the ion-group in the top left junction moves to the middle for a two-qubit gate. The gate is implemented with an external laser beam acting on the two ion-groups.

Reference [193], Steane combines the increasing confidence in the experimental methods for laser controlled trapped-ions with the quantum error correction requirements for a scalable, computationally relevant quantum computer to outline a natural ion-trap model that is experimentally feasible and does not omit any significant technological challenges. The computer is based on the quantum charge coupled device (QCCD) architecture proposed by Kiepinski *et.al.* [112] that proposes a scalable ion-trap design by creating a linear array of ion traps (such as the ones shown in Figure 4.2). The QCCD structure is intended to keep the number of ions chained together in a single trapping region as small as possible, to avoid the technical difficulties of preserving and manipulating large chains of ions. Ions in different interconnected trap arrays can be *ballistically* shuttled (actually physically moved) from trap to trap by changing the trapping potentials of the electrodes. This means that it is possible for any two arbitrarily selected ions to interact, provided that the accumulation of errors that occur during ion shuttling does not destroy the state of the stored qubit in each ion.

A schematic of Steane's ion-trap computer is shown in Figure 4.3. His model of the ion chip is composed of ions trapped between segmented gold electrodes deposited on aluminum substrate [174]. Electrode structures that allows greater scalability (as outlined in Reference [114]) are the planar ion-traps where the ions are trapped above a set of individually addressable electrodes in a plane [47, 95, 178]. The planar traps allow the ions to "float" above the surface of the electrodes, thus allowing the lasers to strike the ions at greater number of angles. The ion-trap electronics under the ion chip in Figure 4.3 allow the control of the trapping voltages from one trap location to another, which, in turn, allows the controlled shuttling of ions from one trap location to another.

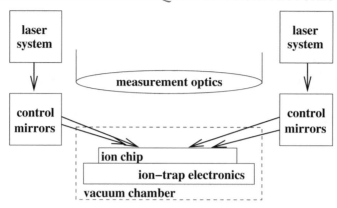

Figure 4.3: Schematic for a scalable ion-trap computer as shown in Reference [193].

The implementation of high-density control electrode interconnects that allow the individual control of millions of trap locations is a significant technological challenge. As described in Reference [114], control voltages can be supplied vertically to the trapping electrodes through the use of via technologies that are currently being made at the densities and dimensions required by an ion-trap computer [50, 173]. Clearly, a major optimization goal for system designers is to create scalable ion-trap geometries that minimize the necessary electronics infrastructure and laser resource distribution.

The laser systems outlined in Figure 4.3 provide the different laser pulses needed for manipulating the ion-qubits, such as qubit preparation, logic gate operations, measurement, and cooling of ions after logic operations or movement. As shown in Figure 4.2, sympathetic ions are used to absorb the accumulated heating from ion movement and gate operations. The sympathetic ions are cooled using laser beams that are also used for initialization of the data ions [30]. Different laser beams with different wavelengths are needed for gate operations. These laser beams must precisely address individual ions for the reliable realization of both single- and two-qubit gates. In addition to the implementation of logic operations, the computer must be capable of reading out qubit states quickly and reliably.

Fault tolerant system designs that rely on error correcting codes require repeated measurements of individual qubits throughout the application execution, which in turn require the implementation of measurement pump lasers that cause state dependent scattering of photons from the ions. The scattered photons are detected by photo-electric light sensors that measure the number of emitted photons. The necessary precision and sheer size of the measurement optics region may force system designers to create ion-trap geometries that divide the ion chip into an ion interaction/storage region and a separate measurement region. Finally, the system-level parameters, control settings, and optimization techniques of the ion-trap computer infrastructure will depend heavily on the choice of ion species used for computation. Making a concrete choice for a general ion-trap computer is difficult at this stage of development since significant tradeoffs exist between different system requirements for different ion types [193].

4.2   ROBUST ERROR CORRECTION AND FAULT-TOLERANT STRUCTURES

The existence of "good" error correcting codes that allow the design of efficient fault-tolerant structures that overcome decoherence is, perhaps, the most critical requirement for a truly useful scalable machine. Due to the high volatility of quantum data, actively stabilizing the system's state via error correction (EC) will be one of the most resource intensive operations that occur during the execution of the quantum algorithm. Unlike classical computation, which relies on the fact that failures are so rare that it is better to take longer for recovery than to spend extra resources for error correction [94], in quantum computing, errors are so frequent that the speed of the error correction method used is critical for the application run-time. Understanding how to design large-scale quantum architectures that employ quantum error correction and quantum fault-tolerance is, perhaps, the most important skill needed from computer architects interested in quantum computing.

Quantum error correction and quantum fault-tolerance constitute a significant field of research [4, 84, 86, 116, 122, 185, 194, 195] that has produced some very powerful quantum error correcting codes analogous to, but fundamentally different from, their classical counterparts. The most important result, as far as scalable quantum computing is concerned, is the Threshold Theorem [4], which says that an *arbitrarily reliable* quantum gate can be implemented using only *imperfect gates*, provided the imperfect gates have failure probability below a certain *threshold* value. This remarkable result is achieved by: (1) using quantum error-correction codes, (2) performing all computations on encoded data, (3) using fault tolerant procedures, and (4) recursively encoding until the desired reliability is obtained. A successful architecture must be carefully designed to minimize the overhead of recursive error correction and be able to accommodate some of the most efficient error correcting codes.

The basic goal of fault-tolerant quantum error correction is to remove errors from an unknown n-qubit state $|\Psi\rangle$ from accumulated decoherence through some sequence of operations. The amount of decoherence that exists can be abstracted as a random unitary error operator that acts on $|\Psi\rangle$. Our assumption is that other types of noise that cannot be described as a random unitary operator are dealt with through different quantum control protocols used in the construction of the physical qubits themselves. Provided that the error operator acts non-trivially on $t < n$ qubits, error correction provides recovery procedures that can correct t errors on a register of n qubits by transferring the errors to a set of ancillary qubits (to avoid direct measurement of the data qubits). After the transfer, the ancillary qubits are discarded or initialized to $|0\rangle$ for reuse.

In the subsequent sections of this chapter, we give a brief overview of quantum error correction and the error correction codes used in the architectures we discuss. The reader may look at reference [149] for a more detailed description of quantum error correction theory. We base our description of quantum error correction on a class of quantum error correcting codes known as Calderbank-

Shor-Steane Codes [43, 44] that allow relatively straightforward quantum computation using the circuit model without the need to decode the encoded states. The detail that we provide for quantum error correction is there to underscore the reasons why error correction is the defining application for quantum architectures; however, if the reader is already familiar with the concept, they may continue reading with Section 4.2.3, which describes the Steane $[\![7, 1, 3]\!]$ quantum error correcting code.

4.2.1   NOISE MODEL ASSUMPTIONS

When modeling errors that can be corrected using fault-tolerant quantum error correction codes, two common noise assumptions are made in both quantum and classical systems: (1) the noise is *stochastic*, where there is an equal probability ϵ of an error occurring in each position; and (2) errors are *uncorrelated* and occur independently of each other. In practice, errors are not always completely uncorrelated; however, any correlations can be minimized through underlying control protocols. Given these noise assumptions, if there are A locations where an error may occur in a classical circuit and an error occurs at each location with probability ϵ, the probability that t errors occur is given by

$$\binom{A}{t}\epsilon^t(1 - \epsilon)^{A-t}, \tag{4.1}$$

which can be understood as the number of possible ways to have t locations that fail and $(A - t)$ locations that do not.

Classically, let a bit be in the state "0" with probability p_0 and the state "1" with probability p_1. After the occurrence of a noise operator which flips the bit with the transition probability ϵ, the bit will be in the state 0 with probability q_0 and the state 1 with probability q_1. Then the evolution of the classical system for each independently occurring noise operation can be modeled as:

$$\begin{bmatrix} q_0 \\ q_1 \end{bmatrix} = \begin{bmatrix} 1 - \epsilon & \epsilon \\ \epsilon & 1 - \epsilon \end{bmatrix} \begin{bmatrix} p_0 \\ p_1 \end{bmatrix}, \ \text{or} \ \vec{q} = E\vec{p}, \tag{4.2}$$

Where E is the matrix of transition probabilities. In quantum computation, the evolution of a quantum system can be modeled in a similar manner. Suppose that we would like to apply the gate U on an n-qubit quantum register $|\Psi\rangle$. Even if the technology allows the state to be completely isolated from the environment, the physical mechanism used to implement the gate will introduce an error (with some probability ϵ) to our original state, due to the fact that the possible unitary operators that can be applied on a state $|\Psi\rangle$ form a continuum. The final state after we apply the gate U can be written as:

$$|\Psi\rangle \ \rightarrow \ \sum_a E_a U |\Psi\rangle \otimes |a\rangle, \tag{4.3}$$

where $|a\rangle$ are states of the environment (unentangled from the data register), and E_a are summed over 2^{2n} possible error operators that act on our state after the gate has been applied. Each error operator is a string of n Pauli matrices $\{I, X, Y, Z\}$ given in Equation 2.15 (and repeated below):

$$I = \begin{bmatrix} 1 & 0 \\ 0 & 1 \end{bmatrix}, \quad X = \begin{bmatrix} 0 & 1 \\ 1 & 0 \end{bmatrix}, \quad Z = \begin{bmatrix} 1 & 0 \\ 0 & -1 \end{bmatrix}, \quad Y = -iZX = \begin{bmatrix} 0 & -we \\ we & 0 \end{bmatrix},$$

For example, after a three-qubit operation on a three-qubit register, the error on the three qubits may be any of the possible combinations of

$$\{I, X, Y, Z\}^{\otimes 3} = \{X \otimes I \otimes I\}, \quad \{Z \otimes I \otimes Y\}, \quad \cdots,$$

where in each n-bit Pauli operator, the i'th entry acts on the i'th qubit (in the subsequent text, the $\oplus$ signs within each n-qubit Pauli operator will be omitted). Noise described by the Pauli matrices is known as *depolarizing noise*. All error correcting codes correct depolarizing noise, and, in general, devices can be engineered such that the dominant noise channel is the depolarizing channel.

Most error correcting protocols rely on the fact that the weight of the n-bit Pauli operators is small, and that the occurrence of highly correlated errors that affect more than one qubit at each step is very rare. However, it is very unlikely that the technologies will allow the complete elimination of correlated errors. The Kane technology [108], for example, stores qubits in the electronic spins of Phosphorous atoms embedded in Silicon. Qubit interactions are controlled via metallic control structures built on the surface of the Silicon substrate. To perform a two-qubit operation, the electron which stores the qubit from one atom is transferred to the other atom. Along the transfer process, the charge fields generated by the control structures interact with the qubit states stored in the data electrons, and, in reality, this fact poses the biggest challenge for physically realizing reliable quantum operations using the Kane technology.

An error on a single qubit happens when a failure occurs during the execution of an operation on that qubit. A failure of the two-qubit CNOT gate can introduce two errors in a quantum circuit, one on the control qubit and one on the target qubit. Based on this assumption, the noise model we use to study the behavior of quantum architectures is as follows:

- Failures are uncorrelated and stochastic. This means that an error on qubit q_i will not result in an error on qubit q_j unless the two qubits are explicitly entangled in the quantum circuit.

- An error on a single qubit can be written as a superposition of the Pauli operators, where a failure at a single timestep takes the density operator of our original system state $|\Psi\rangle$ to:

$$|\Psi\rangle_{new} = (1 - \epsilon)we|\Psi\rangle + \frac{\epsilon}{3}(X|\Psi\rangle + Z|\Psi\rangle + Y|\Psi\rangle)$$

In other words, before the execution of a gate, the qubit undergoes a rotation by X, Z, or Y with probability ϵ, and remains unchanged with probability $(1 - \epsilon)$.

- A two-qubit gate introduces two errors with probability ϵ on the input qubits, equivalent to any of the 15 possible error patterns on two qubits: $\{IX, XI, IZ, ZI, IY, YI, XY, YX, XZ, ZX, ZY, YZ, XX, ZZ, YY\}$, each with probability $\epsilon/15$. Single qubit gates introduce an error with probability $\epsilon/3$ (any of the three possibilities between $\{X, Y, Z\}$). The single-qubit T gate is the only exception to the above, which introduces an error that can be written as a superposition of the X and Z gates.

- Memory and movement are divided into unit operations with a clearly defined failure rate. A particular technology model has a predefined distance that each qubit can travel in the duration of a single gate cycle, with a specific failure rate ϵ that the MOVE gate will introduce an error. Similarly, a memory cycle is equivalent to the qubit staying idle for a single gate cycle, with a specific failure rate ϵ that the qubit will decohere.

- Steane [195] makes the important assumption that qubits participating in a gate at a given cycle undergo only gate noise and not memory noise. Similarly, qubits that move undergo movement noise (introduced by the channel) and not memory noise.

As mentioned earlier, errors are not completely uncorrelated. Initially, the state of a quantum computer is prepared such that it is as independent of the environment as the implementation technology will allow. As the computer state becomes entangled with its surroundings, the amount of entanglement governs how strong the error correlations are between single qubits in the quantum computer. In addition, the application of a logic gate on a qubit also causes an unknown error operator to be applied on the state of the environment — thus the noise at each timestep is shared between the computer system and the environment. If the entanglement between the two systems is small compared to uncorrelated gate failure rates ϵ, correlated errors do not asymptotically affect the scale of reliability achieved due to error correction. The qubit states in the ion-trap technology for example, are affected by phase changes due to the fluctuating global electric and magnetic fields on the ion-trap chip. By fixing a single ion-qubit to be encoded using two physical ions such that

$$|0\rangle \rightarrow |01\rangle; \quad |1\rangle \rightarrow |10\rangle; \quad |+\rangle \rightarrow \frac{1}{\sqrt{2}}(|01\rangle + |10\rangle),$$

(known as a *decoherence free subspace* (DFS) [133]), the correlation of the qubits with the environment can be significantly reduced by protecting them from a phase rotation on both qubits. Any phase error on both qubits will flip the sign of both the encoded $|0\rangle$ and $|1\rangle$ states, which would make the error global and it can be factored out.

An example of non-stochastic errors in the computer are small rotations in each of the qubits, introduced by the classical control mechanisms at each gate. These rotations can be a constant change in the phase or a random rotation at each gate. If the state is randomly rotated by a very small angle θ, then the total angle of rotation after m operations will be approximately $\sqrt{m}\theta$ with probability $m\theta^2$ [195]. As system designers, we must make the assumption that coherent non-stochastic errors will eventually add up to sufficiently large rotations, which can be discretized into a superposition of the Pauli operators as assumed in the incoherent noise model.

4.2.2   ERROR CORRECTION: BASICS AND NOTATION

Perhaps, the simplest way to deal with errors is to detect the presence of errors without correcting them. Errors are detected using *error-detecting codes*, which are employed in classical computation in the transferring of information packets through noisy channels, Arithmetic Logic Units (ALU), and in memory systems. If error detecting codes are computationally inexpensive enough and the classical transmission channel introduces sufficiently small number of errors, then it may be cheaper to retransmit the information packet upon the detection of an error rather than calculating the exact error location.

In general, an error detecting code C is defined by two parameters n and k, where n is the number of bits used to encode a piece of information and k is the minimum bits necessary to represent the information (C is denoted as an $[[n, k]]$ error detecting code).

An example of a common error detecting code is an $[[n, n-1]]$ *parity check* code, which introduces only one bit of overhead. There are many ways to construct parity check codes. One simple way is to count the number of times the bit 1 appears in the original n-bit binary message string. If the number of 1's is even, an extra bit of 0 is appended to the string, otherwise, a 1. For example, the original message codeword 101 has an even number of 1's, so the additional check digit should be 0, changing the codeword to 1010, which is now a codeword in a $[[4, 3]]$ code. A single error on any of the original bits will change the parity of the example codeword from even to odd (note that two errors will not be detected). Thus, upon receiving the codeword, a parity check is done by counting the number of 1's and determining whether the parity bit matches the parity of the received codeword. If the parity bits don't match, then the message is discarded and a request is made to resend the data. Note that the parity check code supports the detection of any odd number of errors. The code provides no information at all about the actual location of any error that has occurred.

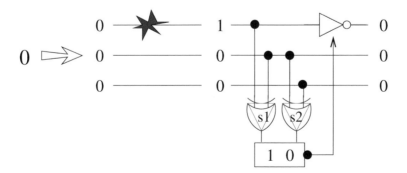

Figure 4.4: Correction procedure with the 3-bit classical repetition code. After the error occurs on the first bit, the syndrome bits ($s1, s2$) are set by measuring the parity between bits (1,2) and (2,3). The correction operation is just a NOT gate on the bit where the error has occurred.

To *correct* errors, it is required that data to be encoded in such a way that both the existence and the location of errors can be identified. One of the simplest error correcting codes is the 3-bit repetition code with codewords "000" and "111". Each bit in the original information bitstring is redundantly encoded three times, where the bit "0" becomes "000" and "1" becomes "111". If an error occurs causing one of the bits to be flipped, a majority vote is taken to determine the location of the error (the three-bit repetition code is an [[3, 1]] code able to correct, at most, one error in any of the codewords). For example, if the received bitstring is "110" and the error probability ϵ is sufficiently low, it is safe to assume that the string encodes the bit "1" and the original value can be easily recovered, or the error can be corrected by flipping the value of the third bit to "1".

The error correction procedure is illustrated in Figure 4.4, where initially the bit "0" is repeated three times as 000 and sent through a noisy channel, where an error on first bit flips its value to 1. The majority vote is taken by simultaneously measuring the parity (we.e. applying the XOR gate) between the first and second bit, and the second and third bit, and the result is stored in the syndrome string (s_1, s_2). In Figure 4.4, the measured syndrome is $(1, 0)$, which indicates that the error is in the first bit. The syndromes $(1, 1)$ and $(0, 1)$ indicate that the error is in the second and third bits, respectively. Clearly, the 3-bit repetition code cannot help us if more than one error occurs. In fact, two errors will cause the error correction to correct the wrong bit, introducing an error in our computation. To detect and correct two errors, majority vote for a 5-bit repetition code can be used (we.e. $0 \rightarrow 00000$, and $1 \rightarrow 11111$). In fact, in order to detect t errors, the distance between valid codewords must be $t + 1$. In order to correct t errors, the distance between valid code words must be at least $2t + 1$.

For a quantum error correcting code, a little more is needed. A logical qubit state is encoded into the state of a number of physical qubits, which are not replicated, but are entangled with one another. In addition, we need to worry about sign errors due to the phase-flip Z operator. An example quantum error correcting code that is similar to the classical repetition codes and can correct both types of error uses 9 physical qubits to encode a single qubit of information, as three blocks of three qubits each [185]:

$$|0\rangle \quad \longrightarrow \quad |\bar{0}\rangle = \frac{1}{\sqrt{8}}(|000\rangle + |111\rangle)(|000\rangle + |111\rangle)(|000\rangle + |111\rangle)$$

$$|1\rangle \quad \longrightarrow \quad |\bar{1}\rangle = \frac{1}{\sqrt{8}}(|000\rangle - |111\rangle)(|000\rangle - |111\rangle)(|000\rangle - |111\rangle) \tag{4.4}$$

Where the logical $|0\rangle$ and $|1\rangle$ states are written as $|\bar{0}\rangle$ and $|\bar{1}\rangle$, and an arbitrary *encoded* one-qubit superposition state can be written as the following:

$$|\bar{\Psi}\rangle \quad = \quad \alpha|\bar{0}\rangle + \beta|\bar{1}\rangle \tag{4.5}$$

Bit-flip errors can be detected and corrected by comparing the values of qubits within blocks, while by comparing the signs of the three blocks phase-flip errors can be detected and corrected. Because all errors are a combination of X and Z errors, this code can correct an arbitrary single-qubit error on any of the 9 qubits used in the encoding.

The circuit that encodes 9 qubits to represent a single encoded qubit as a superposition of the $|\bar{0}\rangle$ and $|\bar{1}\rangle$ states is shown in Figure 4.5, where the arbitrary single qubit $|Q\rangle = \alpha|0\rangle + \beta|1\rangle$ is encoded to $\alpha|\bar{0}\rangle + \beta|\bar{1}\rangle$. The data that needs to encoded and protected is stored in qubit $q1$ as the arbitrary state $|Q\rangle$, which is entangled with eight additional qubits $\{q2 - q9\}$, each initialized to the $|0\rangle$ state. The first two CNOT gates distribute the state of qubit $q1$ into qubits $q4$ and $q7$, similar to the 3-bit repetition code: $|q1, q4, q7\rangle \longrightarrow \alpha|000\rangle + \beta|111\rangle$. The three Hadamard gates transform the three qubit state into:

$$|q1, q4, q7\rangle \quad \longrightarrow \quad \frac{\alpha}{\sqrt{8}}|+++\rangle \quad + \quad \frac{\beta}{\sqrt{8}}|---\rangle, \tag{4.6}$$

where $|+\rangle$ is the familiar $(|0\rangle + |1\rangle)/\sqrt{2}$ state. The state the three qubits are in after the Hadamard gates allows the correction of a phase-flip Z error on any of the three qubits if we compare the signs between qubits $(q1, q4)$ and $(q4, q7)$. To enable the correction of bit-flip errors, the three qubits are encoded with the 3-bit repetition code using the other six qubits $\{q2, q3, q5, q6, q8, q9\}$, where each $|+\rangle$ and $|-\rangle$ become:

$$|+\rangle \quad \longrightarrow \quad \frac{1}{\sqrt{2}}(|000\rangle + |111\rangle)$$
$$|-\rangle \quad \longrightarrow \quad \frac{1}{\sqrt{2}}(|000\rangle - |111\rangle)$$

The result is the encoded arbitrary single qubit state $(\alpha|\bar{0}\rangle + \beta|\bar{1}\rangle)$ as given in Equation 4.4. The method for extracting the syndrome of a bit-flip error in any of the three qubits within each group of three is identical to the classical 3-bit repetition code. Each of the three blocks of three qubits are in the state:

$$|q_1 q_2 q_3\rangle \quad = \quad |000\rangle + |111\rangle, \tag{4.7}$$

where the global phase factor of $\frac{1}{\sqrt{8}}$ has been omitted. The 2-bit syndrome string that would tell us which qubit was flipped can be obtained by performing the parity checks $(q_1 \oplus q_2)$ and $(q_2 \oplus q_3)$. For example, if a bit-flip error on qubit $q3$ occurs:

$$|q_1 q_2 q_3\rangle \quad = \quad |000\rangle + |111\rangle \quad \rightarrow \quad |001\rangle + |110\rangle, \tag{4.8}$$

the syndrome measurement should yield the syndrome bitstring $(0, 1)$. The correction step is then performed by applying an X gate on the flipped qubit. Bit-flip errors are determined in a similar way for the remaining two blocks of three qubits, $|q_4 q_5 q_6\rangle$ and $|q_7 q_8 q_9\rangle$.

The phase-flip Z errors are detected and corrected on any one of the 9 qubits by comparing the signs of the three blocks. If a phase-flip error occurs, for example, on qubit $q6$, then the sign of

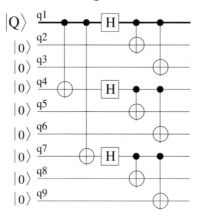

Figure 4.5: Encoding procedure for the quantum 9-bit code. A 3-qubit state is prepared initially that allows for the detection and correction of Z errors. Each of the three qubits is encoded with the quantum 3-bit repetition code to protect against bit-flip errors using 6 additional qubits.

the middle block will be flipped as shown below:

$$|\bar{0}\rangle = \frac{1}{\sqrt{8}}(|000\rangle + |111\rangle)(|000\rangle - |111\rangle)(|000\rangle + |111\rangle)$$

$$|\bar{1}\rangle = \frac{1}{\sqrt{8}}(|000\rangle - |111\rangle)(|000\rangle + |111\rangle)(|000\rangle - |111\rangle) \tag{4.9}$$

Thus, the resulting syndrome string obtained by measuring the parity between the block 1 and block 2 and the parity between block 2 and block 3 should give us the syndrome $(1, 1)$, indicating that there was a Z error in the middle block. Curiously, we can apply the correction on any one of the three qubits in the middle block ($q4, q5$, or $q6$), and the sign will be flipped to the original state. The 9-qubit code is guaranteed to correct any one X or Z error in any of the 9 qubits in the state. It will not correct more than one Z error, but it may correct some higher weight X errors. For example, the error operator "$IIXIIXIIX$" of weight $w = 3$, where there is an X error on qubits $q3, q6$ and $q9$, causes all three errors to be in separate blocks, allowing the 9-qubit to detect and correct the errors. On the other hand, the error operator "$XXIIIIIII$" will cause the first block to correct qubit $q3$, which will be wrong and the entire encoding will be taken out of the codespace, destroying the data that we are trying to protect.

The 9-qubit code is just one example of an error correcting code and is perhaps the simplest truly quantum error correcting code that is capable of correcting both bit-flip and phase-flip errors. Many more quantum error correcting codes are known, where in general a quantum error code C encodes k qubits into the state of n lower-level qubits and can correct errors on up to t qubits. Typically, codes are identified by the three parameters $[[n, k, d]]$, where d is the *code distance* such that $t = (d - 1)/2$. The 9-qubit Shor code can be thought of as a $[[9, 1, 3]]$ code, whose distance d

is equal to 3. A code that corrects any combination of 2 errors in its encoded codewords will have distance equal to 5.

It is not enough, however, to simply store error-free quantum information, we must also have a way to reliably process it for the duration of the algorithm. If a qubit is encoded and protected with some $[[n, k, d]]$ error correcting code, decoding it before processing will prove fatal, for the gates in quantum computation introduce an error with probability ϵ each time a gate is applied. Von Neummann [218] proposed that a classical computer with noisy gates can be made more reliable by performing each gate a number of times and accepting the majority of agreeing gates as the correct gate function. This would require the creation of multiple copies of the data to be sent through the same gate-type, something that cannot be done in quantum computation. The solution is to perform operations on states encoded states. Operations need to be performed *fault-tolerantly* (where more errors are not introduced than it is possible to correct).

A 9-qubit encoded state that forms a single *logical qubit* guarantees protection of the encoded data from any one error which happens with probability ϵ. The data will be lost if more than one uncorrectable errors occur, but if we never decode, higher errors occur with exponentially smaller probability (see Equation 4.1).

In general, performing quantum computation on registers composed of n logical qubits $\{Q_1, Q_2, \cdots Q_n\}$, where the qubits are encoded with clearly defined logical computational states $|\bar{0}\rangle$ and $|\bar{1}\rangle$, is functionally no different than computing with physical qubit registers. A logical gate $\overline{U}$ is constructed from a number of physical gates such that the function of $\overline{U}$ on an arbitrary logical qubit state is the same as the function of a corresponding physical gate U on functionally the same arbitrary physical qubit state. For example, applying the operator "$IIZIIZIIZ$" on an arbitrary 9-qubit logical qubit state encoded with the 9-qubit code will change the sign of each of the three blocks that make up the logical states $|\bar{0}\rangle$ and $|\bar{1}\rangle$, effectively flipping the value of the logical qubit from $|\bar{0}\rangle$ to $|\bar{1}\rangle$, or $|\bar{1}\rangle$ to $|\bar{0}\rangle$. Thus, using the 9-qubit encoding described in this section, the logical bit-flip operator $\overline{X}$ is implemented by applying a Z gate on qubits $q3$, $q6$ and $q9$. Similarly, the 9-qubit operator "$XXXXXXXXX$" (we.e. applying an X gate on all 9 qubits) is equivalent to applying a logical $\overline{Z}$ gate, taking the state $\alpha|\bar{0}\rangle + \beta|\bar{1}\rangle$ to the state $\alpha|\bar{0}\rangle - \beta|\bar{1}\rangle$.

Unfortunately, the implementation of other logical gates is not as straightforward with the 9-qubit code. Therefore, it is important to consider the universal gate implementation circuitry when choosing an error correcting code for a given application. During computation, each logical gate may be followed by a syndrome extraction procedure which would correct any errors (X, Z, or both) that have occurred during the sequence of operations that implement the gate.

There are three major obstacles to overcome when performing error correction on encoded qubit states:

- Quantum states live in a continuous space, identified by the probability amplitudes of the state vector. This means that errors are continuous and, in principle, it should take an infinite number of resources to determine the exact error that has occurred (a single qubit may any combination of bit-flip X errors, phase-flip Z errors, and bit-flip and phase-flip Y errors).

- Measurement destroys the superposition of quantum data, but the only way to extract the error syndrome is by measuring an encoded qubit. Thus, we must indirectly measure the qubit such that its quantum information is not destroyed.

- Quantum data is fundamentally more faulty than classical data. Even if an implementation technology becomes extremely reliable, it may not be possible to generate fewer than 1 error for every 10^8 operations [156] for ion-traps, for example. In addition, quantum data is entangled. This means that quantum error correcting codes must prevent decoherence not only at higher than classical error rates, but must also be designed to limit the effects of entanglement on how errors propagate through the error correcting circuits after each underlying operation. Section 4.2.5 details how quantum fault-tolerance achieved through concatenated quantum error correction can greatly reduce the error rate of quantum operations.

One of the most remarkable characteristics of the theory of quantum error correction (QEC) is that the application of a quantum error correcting code causes errors in the data to be *discretized* [185], thus solving the first obstacle. Unlike classical analog systems, any arbitrary error on one or more qubits may be corrected by correcting a small discrete set of errors: namely X, Z, and the combined X and Z errors. After an arbitrary error, the operator E_i is applied to the *we'*th qubit in the encoded logical qubit, the data state $|\Psi\rangle$ can be written as a superposition of the original state $|\Psi\rangle$, $X_i|\Psi\rangle$, $Z_i|\Psi\rangle$, and $Z_i X_i|\Psi\rangle$. No matter how small the error is, the error syndrome extraction procedure collapses the data state into one of the four elements of the superposition, which can then be corrected by applying either an X, Z, or both. The nice property in error discretization is that extracting the error within a logical qubit can be done simply by extracting a syndrome for X errors and then a syndrome for Z errors, each followed by the corresponding correction operation.

The second obstacle is the inability to measure an encoded qubit directly to extract the error syndrome. Interestingly, this obstacle is not fatal either — however, it does introduce a large auxiliary qubit overhead. In order to measure the error syndrome without collapsing the data, the error information is transferred from the encoded qubit to a number of specially encoded *ancillary* qubits, which are then decoded and measured to reveal the location of the error. Commonly in the QEC codes we discuss here, interaction between the encoded data and the *ancilla* to extract the error syndrome for an $[[n, k, d]]$ code is done in a method known as the *Steane Method* for syndrome extraction [196], which is shown in Figure 4.6.

The Steane method for X and Z syndrome extraction is commonly used for Calderbank-Shor-Steane (CSS) quantum error correcting codes [44]. As shown in Figure 4.6, two sets of n ancilla qubits are encoded using the same error code as the data. To measure X errors, the ancilla is prepared in the logical $|\overline{+}\rangle = \frac{1}{\sqrt{2}}(|\overline{0}\rangle + |\overline{1}\rangle)$ state and a logical CNOT gate is applied between the data block of n qubits as control and the ancilla block as target. A CNOT gate propagates bit-flip errors forward (we.e. control $\rightarrow$ target) — thus the bit-flip X errors from the data block will be transferred to the ancilla. The error and the location of the error can be extracted by measuring the ancilla in the computational basis. To detect and correct phase-flip Z errors, the ancilla is prepared in the logical $|\overline{0}\rangle$ state and is used as the control block during the interaction with the data (Z errors

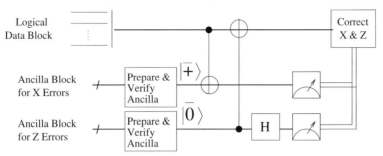

Figure 4.6: Extracting the syndrome and correcting errors using the Steane method for error correction.

propagate backwards in a CNOT gate). Applying a logical Hadamard gate on the ancilla forces the Z errors into bit-flip errors, which can be detected upon measurement in the computational basis.

Measurement of an encoded block of qubits works much the same was as measuring a physical qubit where the state is collapsed to either the logical $|\bar{0}\rangle$ or $|\bar{1}\rangle$ basis states. If the encoded block is intended for a code that corrects up to t errors, measuring a state with any errors of weight $w \leq t$ present will yield the correct measurement (unless some of the measurement gates fail themselves).

The Ancilla Factory Concept: In Figure 4.6, two n-qubit ancillary blocks are shown, one for X errors syndrome and one for Z errors syndrome. What is not shown is the fact that the ancilla blocks themselves must be verified against the presence of X and Z errors after encoding, to ensure that errors created when encoding the ancilla: (1) do not propagate to the data causing errors of higher than correctable weight, or (2) do not cause an incorrect syndrome to be measured. Ancilla blocks can be verified by preparing additional ancilla blocks and using the additional blocks to check each other (much like the error correction process). Constructing better verification structures is critical for the latency of the computation. We could use only a single ancilla block for both X and Z errors and perform the syndrome extraction sequentially by re-preparing the ancilla for each error type. Using a single ancilla increases the error correction time cycle, but reduces physical resource overhead. Alternately, a large number of ancilla blocks can be simultaneously prepared that guarantee that there is always a prepared and verified ancilla block ready for syndrome extraction for both X and Z errors in parallel.

4.2.3   EXAMPLE: THE STEANE [[7, 1, 3]] CODE

For the case studies in the large-scale architecture model presented in this work, we use the Steane [[7, 1, 3]] code [194], which encodes a single logical qubit using 7 physical qubits and can correct up to any single-qubit error. It is based on the classical [[7, 4]] Hamming code, which allows the correction of any single-bit error where the error location is given by the syndrome string that represents the binary numbers between zero and seven. (For example, the syndrome string of "000" denotes no error and the string "010" denotes an error on the second qubit).

The $[\![7, 1, 3]\!]$ quantum code is a member of the CSS family of quantum codes which allow *transversal* logical CNOT gate operations, and whose error correction procedure requires the use of only CNOT and Hadamard gates as shown in Figure 4.6. A logical operator $\overline{U}$ is transversal if its implementation is achieved by applying $\overline{U}$ in parallel to all n encoded physical qubits in a logical qubit block. Furthermore, the $[\![7, 1, 3]\!]$ code is the smallest CSS code that allows transversal implementation of quantum operations, which are members of the *Clifford Group*. The Clifford group is composed of

$$\{H, \text{ CNOT}, X, Z, Y = -iZX, S\}, \tag{4.10}$$

where the S gate is the familiar phase rotation along the $\hat{z}$-axis of a qubit with a phase angle equal to $\phi = \pi/2$ (as defined in Equation 2.14). The T gate (we.e. the other phase rotation gate defined in Equation 2.14) is the only gate that needs to be added to the Clifford group to complete the logically-universal gate set for quantum information processing. The logical construction of the T gate, however, is considerably more complicated with the $[\![7, 1, 3]\!]$ code.

The $|\overline{0}\rangle_L$ logical codeword for the Steane $[\![7, 1, 3]\!]$ code is given by the 7-qubit state:

$$\begin{aligned}|\overline{0}\rangle \quad = \quad & |0000000\rangle + |1111000\rangle + |1100110\rangle + |1010101\rangle \\ & + |0011110\rangle + |0101101\rangle + |0110011\rangle + |1001011\rangle,\end{aligned}$$

where the $|\overline{1}\rangle_L$ state is obtain by applying the logical $\overline{X}$ operator, which is simply 7 one-qubit X gates on each of the 7 qubits in the Steane state. It is straightforward to verify that the action of any of the Clifford group gates transversally on an arbitrary logical qubit state $|\overline{\Psi}\rangle = \alpha|\overline{0}\rangle + \beta|\overline{1}\rangle$ is equivalent to the corresponding physical gate on an arbitrary single-qubit state. In addition, the measurement operation is also transversal. Measuring each of the 7 qubits and calculating the parity of the resulting bitstring will identify correctly if we have measured the logical $|\overline{0}\rangle$ or $|\overline{1}\rangle$ state.

Figure 4.7 shows the circuit used to correct a logical data bit for X errors with the $[\![7, 1, 3]\!]$ code. For the correcting procedure, we use the Steane method, where the steps are:

- First a block of ancilla is prepared in the encoded $|\overline{0}\rangle$ state, as described in [196] and shown in the expanded encoding gate of Figure 4.7. Traditionally the preparation network involves just 9 CNOTgates; however, this would require an additional block of 7 ancilla qubits for the verification, which is applied after the encoding. The circuit shown uses only one ancilla verification bit, and the verification is part of the encoding procedure.

- Second, a transversal Hadamard gate is applied, which places the ancilla in the $|\overline{+}\rangle$ state. The ancilla is then interacted with the data block using a transversal CNOT gate where the data block is the control qubit and the ancilla block is the target qubit.

- The error syndrome is extracted by measuring the ancilla block. If the syndrome is nontrivial (e.g., shows an error), the process is repeated at most three times and terminated if there are two identical syndromes.

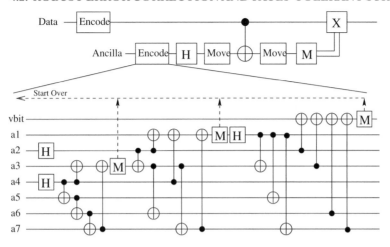

Figure 4.7: Circuit for extracting X-error syndrome and correcting X errors for the Steane $[\![7, 1, 3]\!]$ code using only one verification qubit during the process of preparing the ancilla. All ancilla qubits, a_i, and the verification qubit, $vbit$, are initialized into the state $|0\rangle$ at the start of the circuit.

- Finally, the corrective X gate is applied on the corrupted data bit.

    The same syndrome extraction is repeated for correcting Z errors on the logical data qubit, with the only difference being the flipped control-target blocks for the transversal CNOT gate and the placing of the Hadamard gate after the transversal CNOT(see Figure 4.6). The repetition of the syndrome extraction before the corrective operation is necessary with this encoding procedure because the encoder does not verify the ancilla for Z errors. Any Z errors in the ancilla qubits after preparation are converted to X errors by the transversal H gate, causing the wrong error syndrome. By repeating the syndrome extraction, we ensure that the probability of measuring the wrong syndrome due to Z errors in the encoder is $O(p^2)$, which is a second-order event. Also, the inability to agree on a syndrome after three non-trivial syndrom extractions is taken to mean that more than one error has occurred in the data qubits, which the $[\![7, 1, 3]\!]$ code is unable to correct, deeming the entire computation a failure.

    An alternate encoding network for a logical $|\bar{0}\rangle$ is shown in Figure 4.8. This encoding circuit is given in Reference [201], where three ancillary qubits $\{v1, v2, v3\}$ are required to verify the 7 data qubits $\{q1 \rightarrow q7\}$. The verification ensures that the network is fault tolerant where no single

fault in the encoding network will cause two faults at the output. The authors in Reference [201] make the observation that the only two faults at the output from single faults within the network are on qubit pairs $(q2, q7)$, $(q3, q6)$, and $(q4, q5)$, and thus only three ancillary qubits are required for the verification. Errors are detected in the seven data qubits if the three-bit measurement result of the verification qubits yields an odd parity calculation.

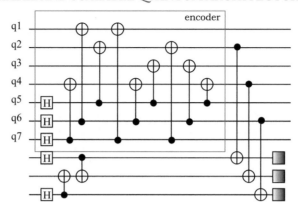

Figure 4.8: The preparation circuit for the logical $|0\rangle_L$ state for the $[\![7, 1, 3]\!]$ code as optimized by Svore et.al. [201].

Generally, it is preferable to choose error correcting codes that allow the implementation of as many transversal logical gates as possible. The fact that the Steane $[\![7, 1, 3]\!]$ code is a CSS code guarantees a transversal CNOT gate between two logical data qubits, and the transversal implementation of the Clifford group gates only makes this code more desirable. The only non-transversal operation for the $[\![7, 1, 3]\!]$ code is the T gate, and while codes exist that allow it to be transversal, the other gates may not be. If the choice of error correction is a compiler-time decision when executing a quantum application, then the compiler must be able to mitigate the cost of the different logical gates needed by the applications. In fact, recent work shows that no code exists where a universal set of gates can be applied [228].

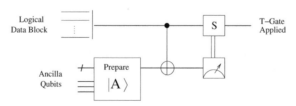

Figure 4.9: T-gate implementation with the $[\![7, 1, 3]\!]$ code.

The T gate implementation with the $[\![7, 1, 3]\!]$ code requires an additional ancillary block, specially prepared such that the concept of *one-bit teleportation* can be used [87]. In general, any arbitrary single-qubit unitary operator U can be implemented using one-bit teleportation, and particularly for the implementation of the T gate, the one-bit teleportation method is shown in Figure 4.9. A 7-qubit $A_{\pi/8}$ ancilla state is prepared using additional ancillary qubits, and interacted with the logical data block to which we want to apply the gate. Since phase information propagates backwards in CNOT gates, the action of the T gate is applied to the logical data block after the CNOT gate with some error. A measurement of the $A_{\pi/8}$ qubit will reveal if the error should be corrected by applying the S gate on the data block.

4.2.4   LOGICAL QUBITS IN QUANTUM COMPUTATION

The *logical qubit tile* is the central architectural abstraction that we introduce as we discuss large-scale quantum architecture design. From a computer architect's perspective, stabilizing an n-qubit quantum state is equivalent to recursively building logical qubit tiles (or blocks) such that the error rate per logical operation, followed by error correction, falls below some desired value that will allow the needed computation to be sustained. Each logical qubit tile at Level L recursion must be crafted in such a way that the failure rate per tile scales as $O(\epsilon^{t+1})$, where ϵ is the failure rate of each Level $(L-1)$ tile used to encode the higher-level qubit. In other words, the physical design and construction of each Level L logical qubit must be *fault-tolerant*.

The efficiency of the design of a fault-tolerant logical qubit tile can depend on several design choices that are not orthogonal: (1) The first and most important design choice is the fault-tolerant error correcting code that best serves the functionality of the qubit tile in relation to the entire processor design. (2) Once a satisfactory code is chosen, the lower-level qubits into the state of which each higher-level qubit is encoded must be arranged as efficiently as possible in order to make the underlying error correcting process more reliable. (3) Another very important design choice is the total amount of lower-level qubits that are allocated for error correction. A number of ancilla blocks may be allocated for a single error correction procedure such that they are prepared in parallel, and we are guaranteed that at least one ancilla block will have passed verification for the extraction of the syndrome. Alternately, we may allocate qubits for only one ancilla block and wait with the syndrome extraction until the ancilla has been prepared and passed verification.

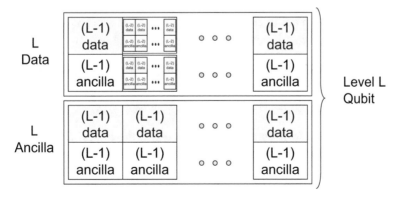

Figure 4.10: Tile-Based Logical Qubit Structure.

A hypothetical schematic of a recursively constructed logical qubit tile is shown in Figure 4.10 without any specific low-level constructions. The figure as a whole shows a logical qubit at Level L, which is composed of a Level L ancilla block and a Level L data block. (The ancilla block is needed for the Steane syndrome extraction method.) Each qubit block at level L is constructed using $2n$ Level $(L - 1)$ blocks, which, in turn, are constructed of Level $(L - 2)$ blocks as shown in the figure. One requirement for the existence of a threshold value (and thus, the ability to increase the reliability with each higher level construction) is that a Level L data block be near the Level L ancilla block used for syndrome extraction [86]. In addition, Figure 4.10 does not show any of the additional ancilla blocks that are needed for extracting X and Z syndromes in parallel, nor for verification of the ancilla blocks used in the syndrome extraction process.

4.2.5 QUANTUM ERROR CORRECTION AND FAULT-TOLERANCE: THE THRESHOLD RESULT

The theory of QEC is powerful and much deeper than we can possibly present here. For it to be truly useful; however, for scalable computationally relevant quantum information processing, there needs to be a way to overcome the exponential spread of errors in an entangled quantum system during the execution of an algorithm. This is especially important because not only are the gates used by the application faulty, but so are the gates involved during error correction.

The formulation of fault-tolerant quantum circuits and the threshold result [4, 85] has made all discussions for scalable, reliable quantum computation with the circuit model possible. The threshold result states that an arbitrarily long quantum computation can be executed with arbitrary reliability using faulty physical gates, provided that the failure rate of each gate is below a certain *accuracy threshold* value. Requirements for the existence of the threshold value are:

- The noise on the quantum hardware occurs independently at each location in a quantum circuit. A location in a quantum circuit is defined as any operation on a qubit such as a gate, or even an idle cycle while the qubit waits for a gate on another qubit to complete. Idle cycles and movement operations on qubits can be abstracted as a WAIT gate and a MOVE gate, respectively. The independence of errors is not a strict requirement, as shown by Aliferis et al. [6].

- Each location in a quantum circuit must introduce an error on the qubit with probability ϵ and it must work perfectly with probability $(1 - \epsilon)$. In other words, the noise is stochastic and the failure probability ϵ depends entirely on the operation type.

- If n qubits are encoded to form a single logical qubit, the logical circuit structures for gates and error correction routines, such as encoding networks and syndrome extraction networks, must be fault-tolerant. A fault-tolerant circuit is a circuit where a single error on any lower-level physical qubit will not spread to $(t + 1)$ or more errors elsewhere in the circuit. (The assumption is that we have an error correcting code capable of correcting, at most, t errors.)

The abstraction for a fault-tolerant CNOT gate is shown in Figure 4.11. The physical CNOT gate is shown in the left-hand-side of the figure where the control and target qubits are both physical qubits. At the encoded logical level, both the control and target are logical qubit structures of n

qubits in an $[\![n, k, d]\!]$ code for the data and the additional ancilla needed for error correction. For the $[\![7, 1, 3]\!]$ code, the logical CNOT gate is transversal and is composed of 7 physical CNOT gates applied in parallel. The error correction step for each logical gate is used as the first error correction step for the next logical gate, so in essence, each gate in a logical circuit is followed by an error correction step. The central assumption is that the number of errors that slip through the logical gate construction network will be corrected by the error correction procedure that follows it, provided that the gate construction and the error correction procedures are constructed fault-tolerantly (where the probability of errors of weight greater than t is a second order event).

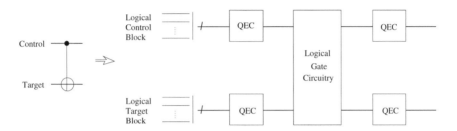

Figure 4.11: Physical → logical gate construction, where a fault-tolerant logical gate is preceded and followed by error correction.

The failure rate of each logical operation that is preceded and followed by error correction can be bounded as

$$\epsilon_1 \leq A\epsilon^{(t+1)}, \tag{4.11}$$

where A is the number of locations, in the gate and the error correction circuits, that cause greater than $(t + 1)$ errors to appear at the output of the circuit. The "1" subscript on ϵ denotes a *single level* of encoding, while ϵ without a subscript denotes the failure rate of a physical gate, which is at level 0 encoding. If a logical qubit is encoded in a block of n qubits, it is possible to encode each of those n qubits again with an m-qubit code to produce an mn encoding. Such recursion, or *concatenation*, of codes can reduce further the logical operation failure rates, provided that the physical failure rates are below the threshold value.

Concatenated error correction introduces an exponential cost with each increasing level of recursion. If each logical qubit block, or each logical line in Figure 4.11, is implemented with an $[\![n, k, d]\!]$ code concatenated L times, then each line consists of at least n^L physical qubits. Figure 4.12 shows the structure of a logical qubit at level L encoding, where Level 1 encoding is defined as the encoding of n *physical* qubits. Encoding once more, for a cost of n^2 physical qubits, we have a logical qubit at Level 2.

Logical circuits composed of logical gates, which themselves are composed of self-similar lower-level logical gates, must obey the same rules of fault-tolerance as the rules for the physical circuit outlined above. An upper bound for the failure rate of a level L logical gate can be defined as:

$$\epsilon_L \;\leq\; A(\epsilon_{L-1})^{(t+1)} \;=\; \frac{1}{A}(A\epsilon)^{(t+1)^L}, \tag{4.12}$$

where ϵ_L is the failure rate of a logical gate at level L encoding. Notice that the "$\leq$" sign will not hold for ϵ_L if the physical component failure rate ϵ is greater than A^{-1}. Therefore, the accuracy threshold value ϵ_{th} for an $[\![n, k, d]\!]$ error correcting code is given as $1/A$, where A is directly affected by the error correcting code used. For a given error correcting code, if the physical component failure rate is below $\epsilon_{th} = 1/A$, we can increase the level of recursion until we reach a desired reliability of computation, or even a reliability that is good enough to sustain computation until the application completes.

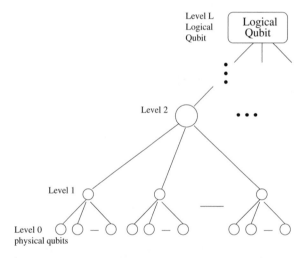

Figure 4.12: The tree structure for a logical qubit using concatenated error correcting codes.

As a system designer, calculating the threshold value for a chosen error correction code will help determine the amount of reliability obtainable with the code at different levels of recursion. The most commonly cited threshold value is $\epsilon_{th} = 10^{-4}$ for the Steane $[\![7, 1, 3]\!]$ code [149]. However, this value assumes noiseless and instantaneous qubit communication in addition to fast and reliable measurement operations.

Gottesman [86] has shown that a threshold value exists in a local setting when qubit communication is considered. In his work, he allows qubits to interact with their nearest neighbors only where movement is performed through successive swapping of the qubit states. The threshold failure rate for a local architecture based on Gottesman's specifications was subsequently computed to be on the order of 10^{-5} [199, 201]. The Steane method for syndrome extraction has allowed a significant simplification in the error correction networks, and thus much higher threshold values have been recently calculated when neither movement nor WAIT gates are considered [168, 195].

In general, any literature published that estimates the threshold value for given quantum circuits has made simplifying assumptions that make the task of calculating the number of fault-locations in these circuits tractable. As system-level designers, however, our main concern is not the exact threshold value, but the design of a fault-tolerant system such that computation can be sustained throughout the application with the minimal number of resources. A qubit at Level L may be encoded using one $[[n, k, d]]$ code, while its lower level qubits may use another. The best way to predict the value of the threshold and the system behavior is through repeated simulations of each component (if exact values of the fault locations A are not available).

4.2.6   THE COST OF QUANTUM ERROR CORRECTION

While quantum computating promises computation that may be exponentially powerful in the number of qubits, coping with decoherence introduces a time and space overhead that is also exponential in the number of qubits and the running time of an algorithm. In this section, we examine this "contest of two exponentials" and outline how to design systems that win this contest and retain the computational advantages of quantum systems.

As the level of concatenation increases when using quantum error correction, the cost of the physical resources and the time it takes to implement single logical gates rises very rapidly. The increase in the physical resource may prove to be the most costly parameter as we recurse since ancillary qubits must be provided for each logical qubit to perform quantum error correction after each logical operation. In addition, level L ancillary qubits must also be error corrected and verified prior to error correcting the data. The increase in computational resources, however, comes with super-exponential decrease in the probability of failure per logical operation. As shown in Equation 4.12, the reliability of logical quantum circuits when using concatenated error correction increases as $(t + 1)^L$. Conversely, the probability of failure ϵ_L per logical operation decreases doubly-exponentially with L for distance 3 quantum codes such as the Steane $[[7, 1, 3]]$ code. Therefore, reaching a desired level of reliability for a given application may only require a few levels of recursion, preserving the exponential improvement over the application's execution on conventional computers.

4.2.7   SCALE-UP IN SYSTEM SIZE DUE TO ERROR CORRECTION

The system size S for a given application can be defined as the product of Q logical qubits and K timesteps [195]. The duration of a timestep is taken to be the time it takes to perform the logical operation, which includes error correcting the n lower level qubits that are encoded in the logical qubit, followed by the time to error correct each logical qubit. The failure rate necessary to achieve a system size $S = KQ$ per logical operation is $\epsilon_{desired} = 1/KQ = 1/S$. A quantum computer with sufficient computational resources may take as many of these resources as necessary for each application to encode data at the desired level of encoding (using a carefully chosen error correcting code) to reach the desired level of reliability for a given system size. For applications with small

KQ parameter, this would leave many qubit resources unused. Another method would be to assume a fixed error correcting code and level of recursion with the hope that the achieved reliability per operation is sufficient for the available set of desired applications. Allowing for a very large system size KQ at all times, however, would be like driving an all-time, four-wheel-drive automobile on a Los Angeles freeway.

Clearly, a lower failure rate could be achieved faster with $[[n, k, d]]$ error correcting codes with $t > 1$ as opposed to the Steane $[[7, 1, 3]]$ code we describe in Section 4.2.3. Such codes, however, use a much higher number of encoded lower level qubits for each logical qubit and the number of locations A that may produce a fault have not been clearly identified, especially when qubit communication is considered within the error correction procedure. In addition, careful studies [195] exist for larger error correcting codes that hint at much more efficient logical circuit structures in terms of resources and latency when $k > 1$. Codes that encode n qubits in $k > 1$ qubits are known as *block codes*, and n is usually quite large. The usefulness of these codes, however, for large-scale quantum architecture is still unclear, as the error correction procedures themselves are very complicated.

To evaluate the expected logical gate failure rate at some level of recursion L for quantum codes where $k = t = 1$, one can use Gottesman's estimate for local architectures [86] shown below

$$\epsilon_L = \frac{1}{Ar^2r^L}(Ar^2\epsilon)^{2^L} = \frac{\epsilon_{th}}{r^L}(\epsilon_{th}^{-1}\epsilon)^{2^L}, \qquad (4.13)$$

where the value for r is the communication distance within Level 1 encoded blocks (defined as the average number of MOVE operations per physical qubit). Equation 4.13 is a rather pessimistic estimate that assumes that the distance the qubits travel before they must be corrected increases exponentially with the recursion level L. While this is true, for sufficiently long distances, the concept of teleportation may be used to change the movement model and allow for lower failure rate ϵ_L estimates.

4.2.8 ERROR CORRECTION SLOWDOWN

From a first look, it seems that the exponential slowdown due to error correction, even with qubit tiles of only a few levels of recursion, is prohibitive when the system size S becomes very large. For some applications, however, the exponential slowdown from error correction is balanced by the exponential speedup offered by the quantum algorithm structure versus its classical counterpart. One such application is Shor's quantum factoring algorithm [184], which is designed to break the widely used RSA public-key cryptosystem. RSA's security is based on the assumption that factoring large integers is very hard, and as the RSA system and cryptography in general have attracted much attention, so has the factoring problem: given a large integer M with two prime factors p and q such that $M = p \times q$, find p and q.

The efforts of many researchers have made solving this classical factoring problem easier for numbers of any size, irrespective of the speed of the hardware. However, factoring is still a very difficult problem. The best classical algorithm known today [40] has complexity of

$$\exp\left((1.923 + o(1))(\log N)^{1/3}(\log\log N)^{2/3}\right)$$

for an N-bit integer. As a basis of comparison, we use the most recent success in factoring using classical computers. A 663-bit number was factored [12] using classical techniques in a process that lasted almost two years on 80 2.2 GHz classical computers.

Shor's quantum factoring algorithm allows factoring of large integers in polynomial time. The algorithm works by using a reduction of the factoring problem to finding the period r of the periodic function $f(x) = a^x \bmod M$, where a is a randomly chosen number co-prime to M, x is an integer such that $M^2 \leq x \leq 2M^2$, and M is the number being factored. The period r divides evenly the number $M' = (p-1) \times (q-1)$ — thus, with some postprocessing, M's factors p and q can be found. The algorithm can be divided into three major steps:

- Modular Exponentiation (quantum part) — computes $f(x)$ in superposition, over all values of x, and by far the most dominant component. The modular exponentiation procedure divides into a series of quantum multiplication steps, which are divided into a series of quantum adders.

- Quantum Fourier Transform (quantum part) — changes the superposition such that the states holding the period of $f(x)$ have the highest probability after measurement.

- Classical Postprocessing — finds the factors p and q from the discovered period r of $f(x)$.

In the performance estimates of the QLA architecture (see Section 7 running Shor's quantum factoring algorithm), the authors have modeled the modular exponentiation routine by using the Draper Quantum Carry Lookahead Adder [69, 214], which has the shortest known expected runtime. Some of the algorithm parameters are shown in Table 4.1.

Modulus	Minimum Operation Failure Rate	Logical Qubits	Timesteps	QFT Timestep Ratio
128	$O(10^{-11})$	37,971	9×10^5	$O(10^{-4})$
512	$O(10^{-12})$	150,771	6×10^6	$O(10^{-4})$
1024	$O(10^{-13})$	301,251	1×10^7	$O(10^{-5})$
2048	$O(10^{-14})$	602,259	3×10^7	$O(10^{-5})$

A plot of the required level of recursion versus the problem size N for factoring an N-bit integer using Shor's algorithm is shown in Figure 4.13(a). The system parameters used are the Steane $[\![7, 1, 3]\!]$ code with ion-trap technology assumptions that are optimistic, but within the fundamental limits of the technology and not out of reach in the future. The details of the architecture are described in Chapter 7. We see that for factoring a 1024-bit (or even a 2048-bit) number, Level 2 recursion with the Steane $[\![7, 1, 3]\!]$ code may be sufficient, given the provided architecture design. The optimistic error rates for the ion-trap technology are almost three orders of magnitude below the existing estimate for the accuracy threshold value of approximately 10^{-5} for the Steane $[\![7, 1, 3]\!]$ code [200].

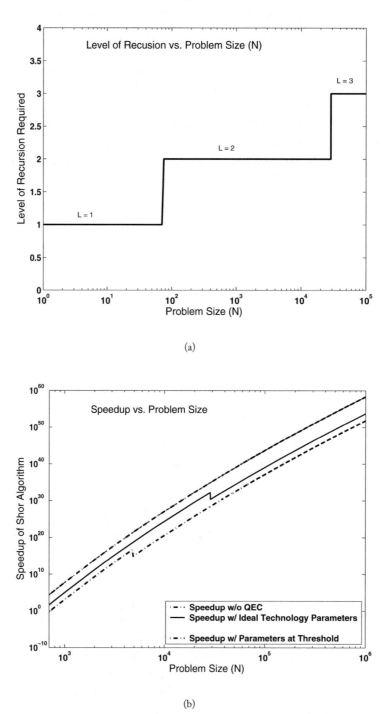

Figure 4.13: Caption on the next page.

Figure 4.13: (a) Required level of recursion for Shor's algorithm as a function of the problem size N defined in the context of an N-bit number that is being factored. (b) Speedup of Shor's algorithm as a function of the problem size N. The top-most line shows the speedup without error correction, the middle line shows the speedup with error correction, but at error parameters approximately three orders of magnitude below the accuracy threshold for the Steane $[[7, 1, 3]]$ code, the bottom line the error parameters are at the threshold value of the $[[7, 1, 3]]$ code. Each "glitch" in the two lower lines is an increase in the level of recursion.

The slowdown due to error correction can be seen in the logarithmic scale plot shown in Figure 4.13(b) where the $\hat{y}$-axis marks the speedup of the quantum algorithm from its classical counterpart. The speedup is calculated as the number of days classically divided by the number of days using quantum computing. The top line is the speedup without error correction. The middle line is the speedup with the optimistic ion-trap parameters, while the bottom line is the speedup with technology error rates at the threshold value of approximately 10^{-5}. Each "blip" on the speedup lines with error correction corresponds to increasing the level of recursion by one unit. The smallest problem size shown is $N = 700$, which requires Level 2 encoding. The same problem size requires Level 3 encoding if the technology parameters are at the threshold value.

As we can see, even with error correction, the exponential speedup is preserved over classical computation. The asymptotic cost of Shor's algorithm is polynomial, and the polynomial cost incurred by the computation is responsible for the deviation of the speedup lines from being truly exponential (note the slight curvature). A physical operation in an ion-trap quantum computer requires on the order of 10 μs — thus, at the physical level, the speedup calculated is based on a KHz quantum computer.

4.3   QUANTUM RESOURCE DISTRIBUTION

Quantum resource distribution is the third major scalability issue that needs to be considered when designing the quantum architecture, an issue that poses a significant challenge for realizing scalable quantum computers. At the lowest level, each qubit is a carrier of quantum information which cannot be cloned and must be physically transported from a source to a destination. This limits us to two methods of communication: (1) either each qubit is a physical transmitter of quantum information, where the qubit itself is physically moved, or (2) operations are applied to transmit the information across a given distance.

Both methods place enormous constraints on the reliability and the speed of quantum data distribution. One method to protect the data from corruption during movement is to repeatedly error correct along the channel, at a cost of additional error correction resources. Another solution is to use the purely quantum concept of teleportation [21] to implement a long-range wire [154], which has been experimentally demonstrated on a very small scale [18, 36, 171]. As described in Section 2.6, teleportation transmits a quantum state between two points without actually sending any quantum data, but rather two bits of classical information for each qubit on both ends. In addition, the coupling of remote atomic qubits (which are well suited for computation) can be achieved through photon interactions [42, 71, 138, 192].

The design and optimization of a quantum architecture to support efficient data communication scalably to arbitrarily large applications will be one of the key areas of contribution for computer architects. We now briefly describe the two types of data distribution: physical qubit movement and movement through teleportation.

4.3.1   PHYSICAL QUBIT MOVEMENT

The CNOT gate has been shown to be the most dominant operation that requires qubit-qubit interaction [200, 201], and hence data movement. There is a large variety of physical qubit communication mechanisms employed by the available technologies to allow two qubits to interact. In fact, the classification of the qubit types heavily depends on the communication mechanisms available for interacting two or more qubits.

Some qubit implementations can be categorized as *flying* qubits (such as photons, which are constantly in motion), and the gates are stationary physical devices that affect the photon-qubits as they "fly" through the gate. Traditionally, photons are sent through fiber optic wires and the main source of decoherence during transport is photon absorption.

Other physical qubit implementations are known as *stationary* qubits, such as the solid-state qubit proposals, which occupy a specific physical space (or a fixed qubit container), where qubit-qubit interactions are limited to nearest neighbor only [58, 108, 136]. The construction of arbitrary one and two-dimensional lattices for logical qubits using stationary qubits is perfectly possible through successfully *swapping* two neighboring qubit states, until two specific qubit states reside in neighboring qubit containers. Gates are applied by an external force, such as electrostatic forces generated from metal gates or lasers that shine on the qubits. The nearest-neighbor communication channels are limited by the reliability of the SWAP operation, (which can be abstracted by applying three successive CNOT gates), although in most nearest-neighbor technology proposals the fundamental qubit-qubit gate is the $\sqrt{swap}$ gate.

Trapped atomic ions that hold the qubit states offer a compromise between "flying" and "stationary" qubits where the ions can be trapped between the segmented electrodes. Lasers can be applied to perform a logic gate at any previously defined interaction region. Two ions interact by ballistically shuttling the ions across the physical layout such that they occupy the same trap[1].

Nearest-neighbor and ballistic qubit communication mechanisms are best suited for implementation of the circuit model for quantum computation as they offer the most straight forward path to reconfigurable quantum logic [226]. From a systems designer's perspective, the two communication models can be indistinguishable: the cost of successive SWAP operations across a swapping

---

[1]An interesting proposal for Josephson Junction qubits support long-distance gates, where any two qubits are allowed to interact without the need to move them; however, the proposal limits the circuit execution to only one gate at a time [131, 215] on a single chip.

channel can be compared to physically moving the ions through a sequence of unit distances in an empty ballistic channel. The challenge for system designers will be to map quantum circuits to physical layouts such that the latency of communication has minimal effect on the latency of the circuit execution. In addition, the physical layout designer must consider that an increased number of SWAP operations, or MOVE operations through a unit length channel, is equivalent to performing random faulty operations on the transported qubit. Thus, great care must be taken to schedule quantum circuits such that programs are optimized not only for latency constraints but reliability constraints.

The error correction procedures for qubits encoded at relatively low levels of concatenation may require an enormous amount of physical qubit movement; however, using clever optimization techniques, it can be possible to reduce the errors data accumulates while in transmission. A significant problem arises when qubits encoded at a relatively high level of recursion must communicate with one another (for example, the execution of a transversal two-qubit gate between two logical qubits at level 3 concatenation). The large increase in the separation between the physical qubits at each additional level of recursion introduces distances that are impossible to traverse physically without a complete loss of data for most known quantum device technologies. In the next section, we describe the concept of quantum teleportation as the means for reliable long-distance communication in quantum architectures.

4.3.2 TELEPORTATION-BASED COMMUNICATION AND QUANTUM REPEATERS

The concept of using teleportation as a long-distance communication channel is illustrated in Figure 4.14, implemented in three stages. The first stages involves the entangling of two qubits into an EPR pair using the network shown in Figure 2.8. The two qubits are then transported through a physical channel where one is moved next to the source qubit and the other to the location where we would like to transport the source qubit. Once the source qubit is interacted with the EPR qubit, the two are measured and the source qubit can be recreated at the destination.

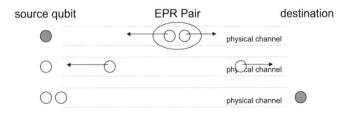

Figure 4.14: Illustration of the different stages of teleportation.

Note that we are still physically moving the entangled EPR qubits. Unlike the source qubit, however, EPR qubits are replaceable. The damaged EPR pairs can be fixed by a process called *entanglement purification* [22, 65], which uses ancillary EPR pairs to distill the good pairs from the bad pairs. The caveat to purification is that the amount of resources increases exponentially with the distance travelled by the EPR pairs, along with the fact that if the EPR pair becomes too corrupted, it may not even lend itself to purification. As the physical distance the EPR pairs must travel approaches the coherence length allowed by the implementation technology, two things occur: (1) the number of additional EPR pairs required for purification of a single EPR pair increases exponentially, and (2) the fidelity of each of the qubits sent through the channel decreases exponentially. For large-scale quantum architectures, we will need to send qubits at distances much larger than the coherence length of the physical channels [60, 140], and it would seem that an enormous amount of resources would be needed to accomplish this.

To allow the successful teleportation of the physical qubits between any two logical qubits in the architecture at an arbitrary distance, we use the fact that entanglement is preserved through teleportation. For example, a logical qubit encoded into the state of 49 physical qubits remains unchanged after some of the physical qubits have been teleported to another region of the architecture. Similarly, teleporting one of the two qubits in an EPR pair does not break the maximum entanglement between the two qubits. The logical qubit interconnect of a quantum architecture can be designed such that EPR pairs can be created between the source and the destination through teleporting the EPR qubits themselves, rather than physically moving them. The EPR pairs, (A1,A2) and (B1,B2), are created and then can be separated apart such that after A2 is teleported over to B2, increasing the span of (A1,A2)'s entanglement. The protocol in Figure 4.15 is known as *entanglement swapping*, where entanglement of EPR pairs is transferred through a channel divided by a number of smaller segments known as *quantum repeaters* [72]. The result is a single EPR pair that spans the entire channel.

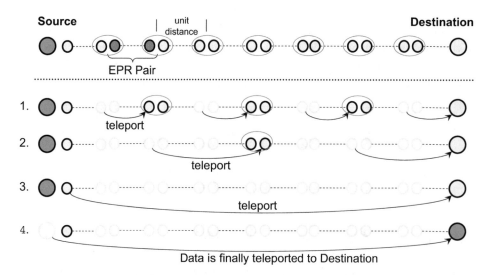

Figure 4.15: Illustration of Entanglement Swapping protocol for teleporting the source qubit marked with "S".

In the entanglement swapping protocol shown in Figure 4.15, EPR pairs only travel to two near-by repeater islands (shaded boxes), where they can be efficiently purified. This protocol allows us to create (expand) a single entangled EPR pair between the source and the destination over the entire channel. As shown in Figure 4.15, each stage of the protocol reduces the number of connecting EPR pairs by half without destroying the connection between the source and the destination. Finally, the source qubit has been teleported to its desired location after a single EPR pair has been prepared to span across the entire connection channel. An EPR pair distributed in such a way between the logical qubits $Q1$ and $Q2$ is required for each of the 49 ion-qubits of qubit $Q1$ such that each of the 49 qubits can be teleported to the computational tile of qubit $Q2$.

Another long-distance communication alternative for logical qubit structures is the nearest neighbor communication protocol, which (at the physical level) is employed by a large number of promising technology implementations where the physical qubit containers are fixed in space [96, 98, 108, 151, 188]. Additionally, recent work by Fowler et al. [77] presents a scalable architecture model for flux qubit quantum computers [96, 151], which at the low-level is dependent on nearest-neighbor interactions. Nearest neighbor communication is analogous to the ion-trap ballistic movement model if each movement "hop" of an ion from one trapping region to the next is replaced with a "swap" operation of the data qubit with another qubit along the direction of movement. Recent threshold studies have shown that the nearest neighbor physical communication mechanism is fault-tolerant [201], and there is no fundamental reason for the physical implementation of quantum architectures (such as the QLA architecture described in Section 7 not to be through nearest neighbor interaction technology. Moreover, circuit synthesis research [179] has shown that nearest neighbor communication preserves the asymptotic depth of quantum circuits.

At the *logical* level, however, a nearest-neighbor-based interconnect will require a number of logical "swap" operations equivalent to the number of logical qubits that separate the source qubit and the destination qubit. To preserve fault-tolerance, each of these "swap" operations must be treated as a transversal logical gate, and thus must be followed by error correction. Given that the duration of a single timestep in the execution of a logical circuit with the QLA architecture is defined by the time to error correct, the nearest neighbor logical interconnect introduces a linear factor slowdown for the application execution equivalent to the average communication distance. For factoring a 1024-bit Shor's algorithm, for example, the largest distance traveled by logical qubits will require 256 swap operations during the modular exponentiation component of the algorithm, and thus 256 additional error correction steps. Choosing teleportation-based interconnect for the QLA architecture allows us to "hide" the temporal cost of communication and decouple it from any logical qubit subroutines. In Section 7.3, we describe in more detail this overlap of communication with computation, which helps alleviate the overhead of additional error correction during movement of logical qubits.

CHAPTER 5

Simulation of Quantum Computation

In addition to the three primary challenges for designing scalable quantum architectures, simulating a quantum architecture may present itself as an additional and significant challenge. Simulation, however, is critical in the ability to verify and validate different architectural designs. In this chapter, we discuss methods available for simulating quantum computers. In particular, we describe a method for simulating the behavior of a large subset of quantum operations by tracking the errors introduced by these operations. We also describe the stabilizer formalism for simulating the behavior of the same subset of operations.

The challenge in simulating quantum architectures comes from the fact that a classical computer cannot efficiently simulate the execution of a quantum program. To see this, one must recognize the fact that quantum information processing is an *extension* of the classical computational model where information is represented as a superposition of classical bitstring states rather than as single classical bitstring. In this sense, the classical computational model is a subset of the larger quantum information processing scheme, and thus, a classical system is unlikely to be able to efficiently simulate a quantum system. However, as the general structure of large-scale quantum computers becomes clearer with each technological advancement, the need to accurately verify and validate the designs of such systems will increase in both urgency and significance.

A unitary operation on an n-qubit quantum register requires $O(2^n)$ operations to simulate or $O(2^n)$ data bitstring entries to store the state of the register. However, even worse for the simulation of quantum computers, is the fact that no quantum computer system is completely isolated from its environment. Quantum systems used for computation are naturally open to external forces which apply the logic gates, and thus, as the system evolves through time, it becomes entangled with the surrounding environment, and unknown forces operating on the environment cause the quantum computer itself to decohere. This means that to accurately track the evolution of a quantum computing system, which is coupled to the environment, one must store more information then is necessary when tracking the superposition state of an isolated quantum register [149].

All existing general purpose simulators [142, 152, 217] incur exponential cost with each additional qubit, and thus simulating even several hundred qubits is unrealistic. Other simulators that impose limits on the entanglement of the system can simulate quantum circuits in polynomial time, as long as the functionality of the circuits satisfies some constraints [209, 219]. A restriction on entanglement, however, is unacceptable for a systems designer who attempts to model error correction, which requires highly entangled qubits for even a single logical codeword.

The situation is not so terrible when one considers the kinds of simulations that system designers need. As a system designer, one does need to track the exact computations performed by a quantum application, but rather the application behavior in the architecture, such as latency, fault-tolerance, and effect on overall system size. There are two types of simulation methods that allow us to model the behavior of quantum circuits using methods that are polynomial in time, but they do not impose any limits on the entanglement produced by the simulated circuit: simulation of error propagation and using the unique *stabilizer* representation of an n-qubit register. Both methods, however, require that the circuits are composed of only the Clifford group gates. Being able to simulate Clifford group operations, however, is sufficient to simulate quantum error correction, which is the bulk of the computational resources [153] during an application execution.

In Section 5.2, we describe in better detail the stabilizer formalism for quantum circuits, where it is used to simulate *stabilizer circuits* directly. Any n-qubit state $|\Psi\rangle$ that can be formed entirely with the Clifford group gates

$$\{H, \text{ CNOT}, X, Z, Y = -iZX, S\}, \tag{5.1}$$

where the qubits must start in the initial state $|0_1 0_2 \ldots 0_n\rangle$, is known as a *stabilizer state*. The stabilizer circuit is the circuit composed of the Clifford group gates that form $|\Psi\rangle$, and any stabilizer state $|\Psi\rangle$ can be described uniquely using only $O(n^2)$ unitary one-qubit Pauli operators $\{I, X, Y, Z\}$. It is a powerful representation for quantum states first published by Gottesman in 1996 [84], where Gottesman provides a description for a very powerful class of error correcting codes known as *stabilizer codes*. The class of CSS codes such as the Steane $[[7, 1, 3]]$ code is a subset of the class of stabilizer codes.

5.1   SIMULATION OF ERROR PROPAGATION

If, as a system designer, one is not concerned with the precise state of the system at a given point in time but rather is concerned with the failure rate of the system as a whole, one can use error propagation to efficiently simulate the behavior of any system component that is implemented using a stabilizer circuit. In addition, long-distance communication mechanisms which, based on teleportation, are also implemented using stabilizer circuits (see Figure 2.2). This means that one can simulate the reliability and efficiency of a long-distance interconnect efficiently on a classical computer using error propagation.

The key to error propagation is that an error on a qubit at any single location of a circuit changes the state of the qubit, which causes any control gates based on that qubit to behave differently. Thus, the error propagates through two-qubit gates and spreads to other qubits as the program progresses. This is why it is absolutely necessary to implement error correcting circuits fault-tolerantly in such a way that an error on any 1 to t qubits will never spread to more than t qubits.

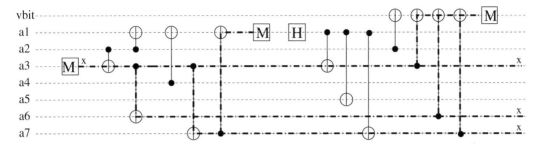

Figure 5.1: X-error propagation. The qubit lines affected by the error are shown in a thicker dot-dashed line. The measurement operations in the middle and the end of the network are designed to yield "1" if error is present and "0", otherwise.

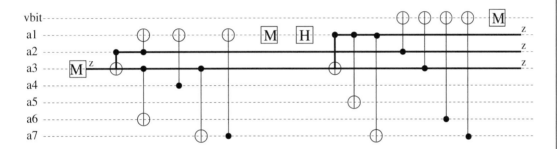

Figure 5.2: Z-error propagation. The qubit lines affected by the error are shown in a thicker solid line. Note that the Z errors are undetected by this network, and a single Z error occurring in the middle of the circuit has caused 3 Z errors in the output.

Consider the simple circuit examples shown in Figures 5.1 and 5.2, which demonstrate the propagation of X and Z errors, respectively. Both networks are carbon copies of the encoding network for the Steane $[[7, 1, 3]]$ code shown in Figure 4.7, but both start at the first measurement operation. Because the measurements are taken in the computational basis, they will detect the states $|0\rangle$ or $|1\rangle$, and the X error will be detected by either measurement gate. Phase-flip errors, on the other hand, slip through the network and have the potential to multiply to more than one error as shown in Figure 5.2. Should more than one Z error slip through the network during the recovery procedure, the syndrome extraction may yield the wrong error syndrome, creating the potential of correcting the wrong data bit (see Section 4.2.3). The fact that any *single* Z error can cause multiple errors and subsequently an incorrect syndrome is the reason why the syndrome extraction is repeated before actual error correction is applied to the data block.

In reality, the effect of errors due to *all* gates can be traced through error propagation simulations. Adding the T gate to the mix of gates whose errors we would like to track would complete the universal set for computation. The problem is that errors introduced by the T gate are described by a superposition of the X and Z gates — thus, we must follow both error paths. With each T gate in a quantum circuit, the number of paths doubles, causing the simulation of the circuit to become intractable very quickly. Most applications such as Shor's algorithm rely heavily on the Toffoli gate described in Section 2.4, which can be composed almost entirely of T gates.

Notice that X and Z errors propagate differently through the two-qubit CNOT gates. Bit-flip errors propagate "forward" i.e., control $\rightarrow$ target), while phase-flip errors propagate "backward" through a CNOT gate. This is easy to see for bit-flip errors since the state of the target bit is flipped depending on the state of the control bit. For phase-flips, we can see this easier if we consider the states of both the input and the target qubit to be $(|0\rangle + |1\rangle) \otimes (|0\rangle + |1\rangle)$. After a Z error on the target qubit, the target qubit state will $(|0\rangle - |1\rangle)$. The application of the CNOT gate puts the system in the state $(|00\rangle - |01\rangle + |11\rangle - |10\rangle)$, which can be written as $(|0\rangle - |1\rangle) \otimes (|0\rangle - |1\rangle)$. Thus, we can see that the Z error has now been propagated to the control qubit as well as the target qubit after the CNOT gate.

Tracking the error behavior of the T gate (which would allow us to simulate universal circuits) through the error propagation methodology makes the complexity of simulation exponential in the number of T gates in the circuit. The T gate introduces a linear combination of phase-flip and bit-flip errors instead of a product of the two. This means that at each gate instance, both paths, the phase-flip path and the bit-flip path, must be followed separately, thus doubling the number of fault paths we are simulating at each T gate.

The extensive set of quantum architecture tools known as QUALE by Balensiefer at al [13, 14] use the error propagation method to verify the fault-tolerant properties of the error correction networks they model. QUALE uses traditional compiler techniques to map quantum circuits onto a realistic physical layout, in order to enable the study of large-scale quantum applications and hardware. The intent of the software tool chain of QUALE is to simplify the development of large-scale quantum applications, where error correction - the most dominant application - is verified through simulating the propagation of errors. Since the noise model is stochastic and errors happen with associated probabilities, Monte-Carlo simulation can be used to find the failure probability of any stabilizer circuit (such as a logical operation as defined in Figure 4.11).

After the network behavior is simulated a sufficient number of times, the simulated failure probability of the entire circuit is the number of registered failures divided by the number of registered successes per trial. A registered failure is any time more errors have propagated to the output of the circuit than the error correction code can correct. If a large-scale quantum computation is composed of a sequence of logical gates like the gate in Figure 4.11, the application is marked as "failed" and is restarted whenever one of two things happen: (1) more errors enter the recovery network than are possible to correct, which would completely change the meaning of the encoded codeword; or (2) either the recovery network or the logical gate circuitry is not fault-tolerant, and it causes a single fault at any location to propagate to more error than the next recovery network can correct.

The drawback of simulating propagation of errors, however, is that the failure probability results are pessimistic when compared to statistical data obtained from other simulation methods.

Without knowing the state of the quantum register, it is impossible to determine what type of fault will actually be a real fault. It is true that any of the Pauli operators are applied with equal probability ϵ, but the phase-flip operator Z, for example, does not affect the $|0\rangle$ state. Thus, simulating propagation of errors sometimes introduces faults on qubits states that are unaffected by the error operator, which makes it effectively a non-error. A logical qubit in the encoded $|\overline{+}\rangle$ state is unaffected by a logical $\overline{X}$ operator; however, if enough X errors have propagated to that logical qubit such that they implement the $\overline{X}$ operator, this will be registered as a logical error and crash the entire application.

5.2 STABILIZER METHOD FOR QUANTUM SIMULATION

Another method for efficiently simulating stabilizer circuits is through the stabilizer formalism [84, 85]. Recall that any arbitrary n-qubit state $|\Psi\rangle$ which can be formed with gates in the Clifford group (provided that all qubits in the register have been initialized to $|0\rangle$) is a stabilizer state. An n-qubit operator U stabilizes the state $|\Psi\rangle$ if U does not change the state: $U|\Psi\rangle = |\Psi\rangle$. The key to the stabilizer formalism's use for the simulation of quantum circuits is the *Gottesman-Knill* theorem, which states that if the n-qubit state $|\Psi\rangle$ is a stabilizer state, then:

- $|\Psi\rangle$ is stabilized by a set of n-qubit operators composed of the Pauli group matrices, given in Equation 2.15.

- The stabilizer group can be generated using $\Omega(n)$ n-qubit Pauli operators, i.e., every stabilizer operator of the state $|\Psi\rangle$ can be written as a product of a small set of stabilizer operators for $|\Psi\rangle$.

- The state $|\Psi\rangle$ is uniquely described by the set of operators that generate all of its stabilizers. These operators are known as the *stabilizer generators* for $|\Psi\rangle$.

The last point states that it is exponentially cheaper to describe a stabilizer state $|\Psi\rangle$ using its stabilizer generators, rather than describing the state explicitly. Consider, for example, the stabilizer generators for some unknown three-qubit state $\{III, ZZI, IZZ, ZIZ\}$, where the i'th Pauli operator in a stabilizer string is understood to act on the i'th qubit only. The first operator III stabilizes anything, because it is just the identity on all three qubits. The second operator ZZI stabilizes the four states $|000\rangle$, $|001\rangle$, $|110\rangle$, and $|111\rangle$. The third operator IZZ stabilizes the states $|000\rangle$, $|100\rangle$, $|011\rangle$, and $|111\rangle$. Finally, the last operator stabilizes the states $|000\rangle$, $|101\rangle$, $|010\rangle$, and $|111\rangle$. Note that common to all four operators are the two states $|000\rangle$ and $|111\rangle$; thus the state stabilized by the generators $\{III, ZZI, IZZ, ZIZ\}$ is $|\Psi\rangle = \frac{1}{\sqrt{2}}(|000\rangle \pm |111\rangle)$.

The total number of classical bits needed to specify an n-qubit stabilizer state $|\Psi\rangle$ is $(2n + 1)$, where the "1" is due to the sign bit, and there are $2n$ X and Z operators to write down. Additionally, Gottesman and Knill showed that unitary operations on the qubits that are part of the Clifford group, such as the CNOT, Hadamard, S-gate, and measurement, take each stabilizer state to a different stabilizer state — thus, the action of these gates can be modeled in only $O(n)$ time. Measurement is slightly more expensive if the outcome is deterministic, where the stabilizer generators can be updated in $O(n^3)$ time. Aaronson and Gottesman have since demonstrated an implementation of a stabilizer-based simulator (known as CHP), where measurement can be updated in $O(n^2)$ time [1].

While we cannot simulate Shor's algorithm exclusively using stabilizer circuits, we can simulate efficiently the most well know class of quantum error correcting codes, namely the stabilizer CSS codes, such as the Steane $[[7, 1, 3]]$ code. In addition, the stabilizer method allows us to simulate some of the most important quantum protocols such as teleportation and superdense coding. In fact, the stabilizer formalism can be used to directly derive encoding and error correcting procedures for stabilizer codes. For example, the set of n-qubit Pauli operators that generates the stabilizers for the encoded logical states $|\bar{0}\rangle$ and $|\bar{1}\rangle$ for the Steane $[[7, 1, 3]]$ is given by the six operators $\{g1, g2, g3, g4, g5, g6\}$, where:

$$\{g1, g2, g3, g4, g5, g6\} = \{XXXXIII, XXIIXXI, XIXIXIX, \\ ZZZZIII, ZZIIZZI, ZIZIZIZ\} \tag{5.2}$$

The reader can verify that applying any of the above operators to the encoded $|\bar{0}\rangle$ and $|\bar{1}\rangle$ states for the $[[7, 1, 3]]$ code (given in Equation 4.11) will not change the two codewords. On the other hand, a Pauli error on any of the 7 qubits will change the stabilizers for the two codewords. Thus, measuring each of the stabilizer generators to determine which ones still stabilize the codeword states is another way of obtaining an error syndrome for the $[[7, 1, 3]]$ code.

Stabilizer circuits can be verified for fault-tolerance and functionality also by using Monte-Carlo simulations The interoperable tool chain QASM-TOOLS developed by Cross et.al. [59] uses the assembly language QASM as an input language to represent and study fault-tolerant quantum circuits by estimating depolarizing noise thresholds using Monte-Carlo simulation, and functionally verifies stabilizer circuits using Aaronson's improved stabilizer simulator CHP. In addition, QASM-TOOLS can find lower bounds for the accuracy threshold of distance 3 codes such as the Steane $[[7, 1, 3]]$ code using general malignant set counting [7], which counts all combinations of locations in a logical gate circuit that cause the network to fail.

CHAPTER 6

Architectural Elements

In this chapter, we discuss the elements of a general quantum architecture. We make the assumption that a quantum architecture will have two primary components:

1. Logical qubits implemented as self-contained computational tiles, allowing gates to be performed directly on the encoded data and containing the necessary error correction resources to correct data immediately following a logical gate execution.

2. Teleportation-based communication channels that may employ the concept of quantum repeaters to allow information transmission across arbitrary regions in the architecture.

Self-contained logical qubit tiles make sense since error correction will be the dominant application executed by a large-scale quantum computer, given that it may be required following each logical gate (see Section 4.2). In fact, error correction must be continuously applied on logical qubits that are not participating in any quantum gates to ensure that the states of the qubits do not decohere. An architecture designer can distinguish between two different types of logical qubit tiles: ones that store logical qubits that participate in logical gate execution; and ones that store logical qubits that simply store quantum information through multiple execution cycles (i.e., through memory gates). Another natural choice would be to just make every logical qubit tile employ the same encoding and error correcting code regardless of wheather the logical qubit stored in each tile undergoes logical gates or memory gates.

Quantum applications (much like classical ones) exhibit natural serialization. By exploiting the limited parallelism at both the application and the physical microarchitecture level of a quantum computer, it is possible to reduce the area requirement while improving performance by designing logical qubit tiles optimized for computation and logical qubit tiles optimized for storage of quantum data. The key observation here is that when the amount of available parallelism in quantum applications is considered, we discover that much is to be gained by limiting computation to a specifically designated region. The remaining area can be optimized for storage of quantum data. A good example for the benefit of specialization in quantum applications is the Draper carry-lookahead quantum adder [69], which forms a basic component of Shor's quantum factoring algorithm [184]. Figure 6.1 shows that providing unlimited computational resources for a 64-bit adder does not offer a performance benefit over limiting the computation to 15 locations. If there are twice as many ancillary qubits as there are data qubits, then by providing only 15 compute locations instead of 64, we can reduce the area consumed by each adder by approximately *half* and yet have no change in performance.

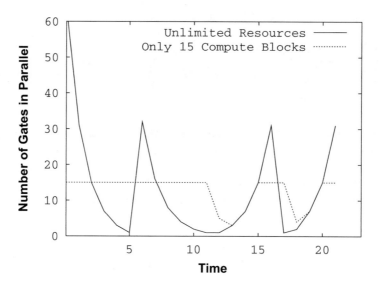

Figure 6.1: The amount of parallelism that can be extracted from a 64-qubit adder when resources are unlimited, and when the number of gates per cycle are limited.

The high-level specialized architecture model we describe here is shown in Figure 6.2. The model is constructed from a collection of specialized architectural elements much like a classical architecture, independent of the physical implementation technology. Each element is composed of a number of "tiles", where each tile represents one or more logical qubits (composed of a number of physical qubits encoded using a pre-specified error correcting code). The shaded tiles in Figure 6.2 represent logical data qubits, and the clear tiles are logical ancilla blocks (used for error correction at the highest level of recursion).

The only logical operation employed by any specific quantum memory would be a WAIT gate (we.e., memory gate), which is simply doing nothing. Assuming that the implementation technology allows a WAIT gate to be considerably more reliable than other gates, the overall computer area can be reduced by changing the ratio of logical data blocks to logical ancilla blocks in the memory and computational regions. For example, the main function of the memory tile shown in Figure 6.2 is to store and error correct encoded data qubits. As the number of ancillary qubits used to error correct a given set of logical data qubits decreases, so does the interval between two successive error correction operations for each logical qubit. This is acceptable in memory tiles as long as the data qubits do not wait too long between error correction steps. On the other hand, error correction is needed after the execution of each gate in the computation region, and thus more ancillary qubits will ensure faster computational times at the expense of more physical resources. In addition, we may be able to combine area savings with improved performance, by defining a specialized *compute code* (CC) used in the processing elements, and *memory code* (MC) used for storing data.

The introduction of different encodings used by tiles that allow computation and tiles that only store qubits will require a complex transfer network between the different tiles, and the data must not be decoded in the transfer process. As we will see later in this section, the transfer network is slow since it is composed of a number of gates and measurement operations on the encoded data, each followed by error correction.

Figure 6.2 shows an additional *cache* region used to buffer data encoded with the computational code, after it is transferred from memory. In some ways, the memory hierarchy we describe in this section is a *code hierarchy*, where the hierarchical structure is needed to overcome the latency differences between state stabilization and code transfer from one encoding to another. The structure and optimization of the hierarchy is perhaps the most complex component of the architecture, as it provides the transition operations necessary to take data encoded in the highest level of the hierarchy to the encoding needed for computation, without delaying the algorithm execution.

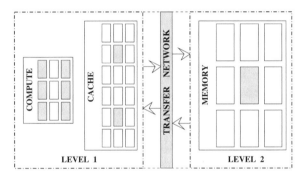

Figure 6.2: High-level view of a specialized quantum architecture.

6.1 QUANTUM PROCESSING ELEMENTS (PE'S)

All logical quantum operations take place in the processing element (PE) tiles. A schematic of a hypothetical PE tile is shown in Figure 6.3. When a logical qubit is teleported to an available PE it is stored in either one of the two *accumulators*, and it is encoded with the compute code (CC), where the CC is chosen to be fast and relatively inexpensive in the number of physical qubits needed for encoding and error correcting a single logical qubit. Error correction is performed before and after the application of a single logical gate on the data stored in any of the two accumulators, using the closer of the two ancillary blocks. The logical qubit may be found waiting in the *quantum cache* encoded with the same CC, or is teleported directly from the main memory if there is an available accumulator in some PE unit. A two qubit gate requires the use of both accumulators, since the physical qubits of each of the two participating logical qubits must interact with one another. There are enough CC ancilla provided to correct both logical qubits in each of the accumulators. In reality, the lines between the different regions in each PE are not as clear as drawn in Figure 6.3. For example, in the ion-trap technology, the execution of a two-qubit gate with the Steane $[[7, 1, 3]]$ code will require 49 pairs of ions to be placed in the same trap. Thus, both accumulators can be constructed by having 49 traps that allow physical two-qubits gates to be executed.

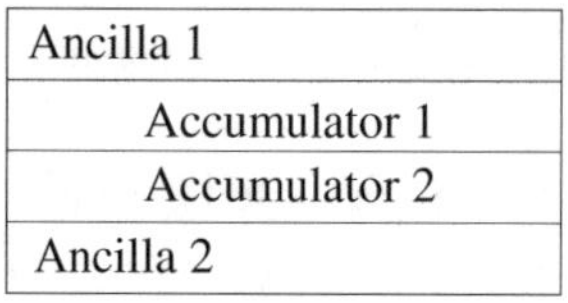

Ancilla 1
Accumulator 1
Accumulator 2
Ancilla 2

Figure 6.3: Hypothetical schematic of a Processing Element (PE) tile.

Gates acting on logical qubits must preserve fault-tolerance, so that a single error on any of the lower-level logical qubits will *not* spread to more lower-level qubits than the CC can correct. The gates act on logical qubits without decoding the states, thus a compiler optimizing the fault-tolerant structure of each gate must have clearly defined transformation rules that preserve fault-tolerance. The best CC's are ones that: 1) Use very little physical qubit resource overhead; and 2) Allow "easy" fault-tolerant gate implementation. Good candidates for CC codes are the Steane $[[7, 1, 3]]$ code, or the newly optimized Bacon-Shor $[[9, 1, 3]]$ code [7, 11, 164]. The Bacon-Shor $[[9, 1, 3]]$ code is based on the well known Shor 9-bit code [185] and allows very fast and efficient error correction routines. However, as shown in Section 4.2.3, the T gate is more difficult to implement. It requires the interaction of the logical qubit with a specially prepared encoded $A_{\pi/8}$ ancilla (also in CC), making the T gate essentially a two-qubit gate [87]. Many of the tiles in the processing region must be used to prepare the $A_{\pi/8}$ logical qubits; Thus, when a T gate is executed the logical qubit and a ready $A_{\pi/8}$ qubit are teleported to two accumulators in an empty PE.

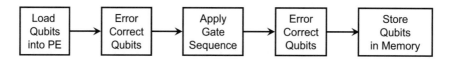

Figure 6.4: The five stages for an instruction execution from the perspective of a processing element.

From the perspective of each PE, an instruction is executed through 5 stages shown in Figure 6.4: 1) The logical qubits are loaded into an available PE; 2) The logical qubits are error corrected; 3) The gate implementation sequence is applied on the logical qubits; 4) Error correction is applied again; and finally, 5) The logical qubits are sent to an available cache address.

6.2   QUANTUM MEMORY HIERARCHY

Classical memory hierarchies optimize for speed, given technologies of differing performance and cost (for example, SRAM and DRAM). In this section, we discuss a *quantum memory hierarchy* that optimizes, instead, for error correction codes, which can either facilitate computation or improve storage density. The quantum memory hierarchy exists to provide the reliability necessary to fault-tolerantly encode and store quantum data for the duration of a given application. The lowest level structures of the hierarchy are designed to meet the speed and efficiency requirements of the processing elements by accepting and storing differently encoded data residing in the higher levels of the hierarchy.

The Steane $[[7, 1, 3]]$ error correcting code is well suited for computation since most operations can be implemented transversally and the code is relatively small. For memory, however, it is worth considering more efficient block $[[n, k, d]]$ codes that correct more than one error and the $[[n, k, d]]$ code scale-up ratio (defined as the number of logical qubits divided by the number of physical qubits) is better than $1/7$. $[[n, k, d]]$ codes where $k > 1$ are known as block codes. The caveat is that using different CC and MC codes calls for a considerably more complex transfer network when transporting a logical qubit resting in memory to an accumulator in the processing region. The transfer of a logical qubit to a different location with a different encoding state code cannot be implemented using just the straightforward repeater-based interconnect used within each architecture region. The transfer network must provide for fast and efficient conversion between the CC and MC codes, as well as exploit temporal and spatial locality to effectively cache data in the CC code. The concept of the memory hierarchy is illustrated in Figure 6.5.

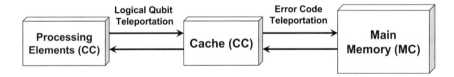

Figure 6.5: Quantum Memory Hierarchy High-Level Concept.

The transfer region between memory and computation (including the cache) is one of the most interesting components of the memory hierarchy. This region *teleports* data encoded in code C1 to a second code C2 without decoding (decoding the data at any time of the computation leaves individual qubits vulnerable to errors). Figure 6.6 illustrates this concept. The transfer network teleports the data in C1 to C2, where C1 and C2 may be any two error correcting codes. The code teleportation procedure works much the same way as standard data teleportation that is used for communication. An entangled ancillary pair is prepared first between C1 and C2 through the use of a multi-qubit cat-state (we.e. $(00...0 + 11...1)$). The data qubit interacts with the equivalently encoded ancillary qubit through a CNOT gate, and the two are measured. Following the measurement the state of the data is recreated at the C2 encoded ancillary qubit. This process is required every time we transfer a qubit between memory and the cache. The most important property of the transfer network is that C1 and C2 need not be two different codes, but can be the same code at different levels of recursion (e.g., CC can be level 1 $[[7, 1, 3]]$ code and MC can be level 2 $[[7, 1, 3]]$ code).

In the quantum architecture organization of Figure 6.2, every region (the processing elements, the cache, the main memory, and the transfer region) is composed of logical qubit tiles interconnected by the programmable teleportation-based bus lines. When a cache hit occurs, the classical control

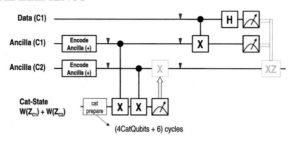

Figure 6.6: Code teleportation network from code 1 (C1) to code 2 (C2); C1 and C2 can even be the same error correcting code, but different levels of encoding. The solid triangles denote an error correction step.

resources are focused on the teleportation channels that connect the cache and the PE, and second priority is given to transferring additional qubits to the cache. This, however, is a scheduling decision and currently no true schedulers exist for large-scale quantum computation.

The next stage of the architecture development is to implement efficient simulators that will allow us to fully exploit and parameterize the architectural design. Parameterizations of the cache read time, memory access time, operation times, and qubit failure rates are not only functions of our error corrections choices, but also functions of interconnect design and the structure of the architectural elements. Some important questions we need to answer involve the error correction choices, cache replacement rules, and the availability of classical resources.

Error Correction Choice: A key decision is to determine the error correction codes for the main memory (MC) and the computation (CC) tiles. Much depends on the parameters and the properties of an error correcting code: the time of execution for logical operations, the size of each tile, the size of the application being executed, and most importantly the coupling between communication and computation at the high-level. In addition, the data communication patterns of different error correcting codes differ wildly and may affect the efficiency of the code itself. For example, the Bacon-Shor $[[9, 1, 3]]$ code is a recently optimized version of the $[[9, 1, 3]]$ code (described in Section 4.2.2) that needs almost no physical qubit movement between the two-qubit physical gates during encoded state preparation and error correction procedures at the first level of encoding. Perhaps, the Bacon-Shor $[[9, 1, 3]]$ code coupled with a different smaller code at the next level of encoding may offer a much more efficient and reliable computational tile than using the same CC code from one level of encoding to the next.

Cache Replacement Rules: It is important to understand clearly the replacement rules when the cache is full. The application being executed is known in advance, so our compiler will be able to schedule the operations and the cache usage statically at compile time, however, preparation of the interconnect channels on demand and time to error correct can only be predicted with a limited accuracy. There is always a certain probability of failure, which leads to stalling.

Code Conversion Choices: When is the best time to convert from CC to MC? The cache can perform memory correction but with limited classical resources, thus qubits should not stay in CC space for too long. To optimize for limited classical resources such as laser FANOUT and control circuitry, classical schedulers must reach a careful balance between the cost of code teleportation between different regions in the architecture.

Classical Resource Availability: What are the available classical resources? In our previous work [140] we have assumed unlimited classical control signals, which take the form of lasers for ion traps. If we have a very small number of lasers available, however, the replacement rules in the cache and the PE units will become extremely important. Currently, the replacement rule is to send a qubit directly to memory when the computation is finished and only send it to the cache if it is needed before the cache storage failure rates. Sending it to memory may prove advantageous because it is designed to store qubits for long periods of time with very small number of laser resources. A careful balance must be reached, however, between the cost of code transfer from memory and memory storage.

6.3   QUANTUM ADDRESSING SCHEME FOR CLASSICAL MEMORY

The separation between memory and compute regions discussed so far is a system-level separation that provides a computer architect with various knobs to turn when optimizing an architecture design for specific quantum applications. A different and interesting separation between memory and computation is offered by the implementation of the quantum search algorithm, known as *Grover's Algorithm* for searching an unsorted data-base of N entries [89]. While classically, the search would take $O(N)$ operations, quantum mechanically the cost is $O(\sqrt{N})$. A naive classical architecture for searching a database is to store all data entries into a long-term memory unit and perform a maximum of N LOAD operations from the memory to the processor for each entry in the database. The freshly loaded entry string is then compared to a solution string stored in the processor.

The main engine for the quantum searching algorithm is the *oracle* operator O whose action can be written as:

$$|x\rangle \rightarrow (-1)^{f(x)}|x\rangle, \tag{6.1}$$

where x is the index register which points to the data entry in the database. The input x into the search function f returns 1 if x is a solution to the search problem and 0 otherwise. The index register $|x\rangle$ is composed of $\log N$ qubits, where each bitstring state $|x_i\rangle$ in the superposition indexes a data entry. Thus, the function of the oracle is to flip the sign of the index register if a solution is found. Along with the n-qubit index register (where $n = \log N$), the processing unit of the search algorithm involves an l-qubit register to hold an l-bit data entry initialized to $|0\rangle$ and an l-qubit register that stores the solution string.

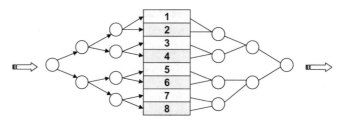

Figure 6.7: Schematic for classical memory that is addressed with qubits. The figure illustrates the concept with a 3-qubit address memory of 8 data entries. Each circle represents an ancillary qubit used as a switch to route the input index register to the correct data entry.

What makes the architecture of the quantum search implementation interesting is that data can be stored in quantum bits that only take the values of 0 or 1, and thus reliable storage of the database is of a smaller concern. For the polynomial speedup to be achieved, an N-entry memory must be addressed quantum mechanically by $\log N$ qubits [149]. Figure 6.7 illustrates the concept with a 3-qubit address memory of 8 data entries. Each circle represents an ancillary qubit used as a switch to route the input index register to the correct data entry. Each of the data register qubits is routed to the corresponding entries in the memory based on the state of each qubit-switch, which is determined by the index register in the processor. The data register qubits enter at the left and exit at the right of Figure 6.7. If a particular switch is in the superposition state $\frac{1}{\sqrt{2}}(|0\rangle + |1\rangle)$, then the data qubit is routed in both directions. In this manner the LOAD operation returns a superposition of data entries that can be compared with the l-qubit register that stores the solution string. To match one of data strings from memory with the solution string stored by the l-qubit solution requires $O(\sqrt{N})$ LOAD operations.

In reality, the searching of an unsorted database quantum mechanically is not more efficient than storing and searching the database classically. The quantum addressed classical memory requires $O(N \log N)$ ancillary qubit switches, in addition to the operations overhead once the data is loaded into memory. Should error correction be needed for storing and searching through large databases, the modestly polynomial improvement over classical searching can be overwhelmed by the slowdown due to error correction. However, should qubits become as easily and cheaply implemented as classical logic transistors are, then it may become interesting to implement quantum addressable memories. One use of a quantum addressing scheme is prefetching classical registers in the classical memory hierarchy. In a single clockstep, the quantum address will be able to fetch an entire superposition of classical registers at any location of the memory.

6.4 ERROR CORRECTION AND QUANTUM ARCHITECTURE DESIGN

Quantum computer architectures make heavy use of quantum error correction to guarantee reliable operation from a comparatively unreliable set of primitives. Because error correction is such an integral part of the design of a quantum system, quantum architects must design their systems around the principles of quantum error correction. In the next chapter, we will look at a few notable quantum architectures whose structure and organization is fundamentally influenced by error correction (in particular, the abstraction of a logical qubit).

Before this, however, we will first look at a potpourri of work where the effect of different error correction techniques and optimizations (gauged via mathematical models) is taken into account to attempt to formulate more effective quantum architectures. While certainly not exhaustive, these examples help illustrate the kinds of things we can do to improve an architecture when we take error correction into mind.

6.4.1 EFFECTS OF ANCILLA PREPARATION AND LAYOUT

Kreger-Stickles et al. [125] presents a domain space exploration of the architecture of an ion-trap quantum computer. This exploration consists of a performance evaluation of several variants of an architecture utilizing an error correction scheme based on the Steane [[7, 1, 3]] code on a Òsea of trapsÓ layout. Each of these variants have different combinations of ancilla preparation technique and ancilla/verification qubit layout. Potential fault point counting (where all potential faults in the system are used in conjunction with component operation fidelities to determine the system fidelity) is utilized in the evaluation of these test cases.

The first case study, presented by Kreger-Stickles et al., concerns two algorithms for preparing and utilizing ancilla qubits. Ancilla preparation techniques using 7-qubits and 4-qubits are considered, as well as a compact version of the 7-qubit technique (which requires fewer verification qubits) and a nearest neighbor version of the 4-qubit technique (where all preparation is done via interactions with neighboring qubits on a linear array). Three hardware layouts are considered for these four ancilla preparation techniques:

- **Vector:** The ancilla bits are arranged in a linear series of traps that are adjacent to unallocated traps for communication and routing. (Figure 6.8(b))

- **Linear:** A set of trapping regions is arranged linearly; since arbitrary bits cannot be routed to one another in this design this is only suited to nearest neighbor designs. (Figure 6.8(c))

- **Folded:** The ancilla and verification qubits are folded around a crossbar. While this technique results in a larger size, this layout reduces the latency and communication requirements. (Figure 6.8(d))

Two scheduling layouts are considered for each of these preparation techniques and hardware layouts: **early verification** and **late verification**. Qubits are verified as soon as possible in early

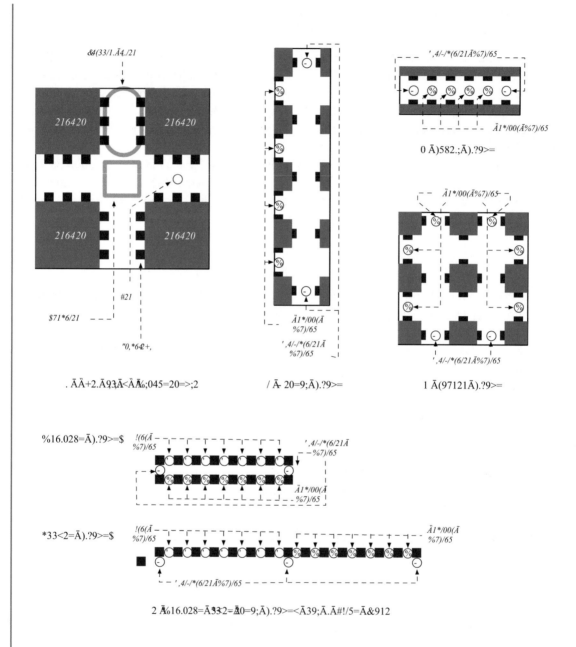

Figure 6.8: a) Sea of traps ion-trap layout b) Vector layout c) Linear layout d) Folded layout e) Adjacent and offest vector layouts for a 7-bit code.

verification, whereas qubits are verified only when the ancilla qubits are next to the data qubits they will be interacted with in late verification.

From computations on circuit run time and potential fault points, Kreger-Stickles et al. concludes that the 4-bit linear nearest neighbor techniques perform very well in both verification scenarios compared to the other techniques. Additionally, computations on the number potential faults in a given communication scenario demonstrate a significantly higher incidence of storage related potential faults for the 7-qubit code in comparison to the 4-qubit code, which also makes the 4-qubit code more attractive for a system with multiple levels of recursive error correction.

The consequences of ancilla preparation and placement choices can also be analyzed in the context of level 1 operations. Before doing this, the authors introduce two more design attributes into the mix: whether the ancilla qubits are arranged adjacent to the data qubits or offset from the qubits. Figure 6.8e shows an example of these layouts for a 7-bit vector arrangement. Simulations of a small level 1 operation on the resulting design permutations show that the offset linear nearest neighbor 4-qubit codes deliver the fastest performance, smallest size, and highest reliability of all the designs tested.

6.4.2 OPTIMIZING ERROR CORRECTION ALONG CRITICAL PATHS

Because error correction is such a common operation in a fault tolerant datapath, optimizing where and when error correction occurs makes it possible to build a faster and more efficient circuit. Whitney et al. [220] present such a technique for optimizing quantum error correction in an architecture designed to implement Shor's algorithm.

It can be assumed that qubits are interacted with other "dirty" qubits (which may have experienced an error) are more likely to accumulate errors themselves. Therefore, that performing error correction only after a qubit has undergone several "dirty" interactions may be more efficient than error correcting after every single operation. However, removing error correction in this manner can increase the probability the circuit fails. Such an optimization requires us to choose between circuit size and reliability.

The basic principle behind this optimization strategy is illustrated in Figure 6.9(a). Whitney et al. [220] define a metric EDist for each qubit to choose when to perform error correction. EDist is updated every time an operation is performed on a qubit. To take into account the errors accumulated in the other qubits involved in an operation, a qubit's EDist is set to the maximum EDist of all the qubits involved in the operation plus one (which takes into account the operation itself). Once EDist reaches a certain value, error correction is performed and EDist is reset. Figures 6.9(b)-(d) show an example of error correction using EDist.

Whitney et al. demonstrate that optimizing an unmapped 1000-bit random circuit in this manner provides very promising results. Performing error correction with EDist = 6 reduces the success probability of the circuit from 0.987 (in the unoptimized case) to 0.956 while reducing the operation count almost an order of magnitude from 3,105,611 to 366,411. Performing the error correction with EDist = 9 reduces the success probability of the circuit more significantly to 0.87

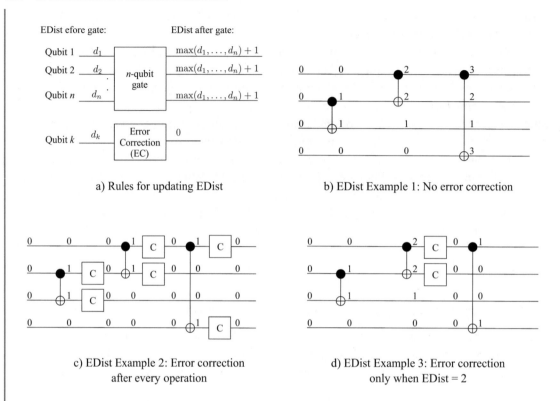

a) Rules for updating EDist

b) EDist Example 1: No error correction

c) EDist Example 2: Error correction
after every operation

d) EDist Example 3: Error correction
only when EDist = 2

Figure 6.9: ECC Optimization

while reducing the operation count by yet another order of magnitude to 24,011. This optimization can also reduce the latency of the circuit, as fewer error correction steps can translate to faster execution time.

CHAPTER 7

Case Study: The Quantum Logic Array Architecture

The chapter describes in detail the design and analysis of the Quantum Logic Array (QLA) architecture, which is a homogeneous, tile-based quantum architecture whose low-level elements are modeled using the ion-trap technology characteristics. Our hope is that by going in detail through the design decisions made for the QLA architecture, the reader will get a better sense in the role computer architectures can play in quantum architecture design. The QLA architecture model is intended to efficiently overcome the primary scalability issues described in Chapter 4, enabling substantial performance improvements critical to supporting full-scale applications such as computationally relevant instances Shor's factoring algorithm.

7.1 QLA ARCHITECTURE OVERVIEW

The QLA quantum computing system, as shown in Figure 7.1, is a homogeneous array of logical qubits implemented as self-contained computational tiles, including all necessary error correction resources, and connected using the teleportation-based communication channels that utilize the concept of quantum repeaters (as discussed in Section 4.3.2).

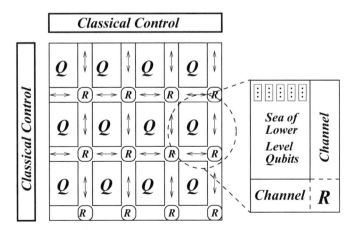

Figure 7.1: High-Level view of the QLA Architecture.

Operation	Time μs now(future)	Failure Rate now(future)
Single Gate	1 (1)	10^{-4} (10^{-8})
Double Gate	10 (10)	0.03 (10^{-7})
Measure	200 (10)	0.01 (10^{-8})
Movement	20 (10)	0.005 (5×10^{-8})/μm
Split	200 (0.1)	
Cooling	200 (0.1)	
Memory time	10 to 100 sec	
Trap Size	$\sim$ 200 (1 − 5) μm	

At the lowest level, the QLA architectures employ the trapped-ion technology. Figure 7.2 demonstrates the abstraction of the physical ion-trap layout, which can be represented as a collection of trapping regions connected together through shared junctions. A fundamental time-step, or a clock cycle, in an ion-trap computer can be defined as any physical operation (one-bit or two-bit) on a single ion-qubit, a basic move operation from one trapping region to another, and measurement.

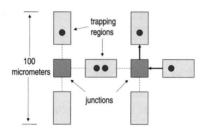

Figure 7.2: The abstraction of the ion-trap layout in the design of the QLA architecture.

Table 7.1 summarizes current experimental parameters and corresponding optimistic parameters for ion-traps. In this chapter, we take each *clock cycle* (i.e., a fundamental time-step) to have a duration of 10 μs, failure rates are 10^{-8} for single-qubit operations and measurement, 10^{-7} for CNOT gates [156], and 10^{-6} per fundamental move operation[1]. Trap sizes are taken to be 5μm each [223] and on the order of 10 electrodes per trapping region [95], which gives a trapping region dimension (including the junction) of 50μm. The parameters chosen for this example are optimistic compared to [13] and [215]. Both of those papers use near term parameters, which are useful for building a 100 bit prototype, but not good enough for building a scalable quantum computer that can factor 1024-bit numbers using Shor's algorithm. Based on the quantum computing ARDA roadmap [222], it seems justified to aggressive (but still realistic) parameters when looking 10 − 15 years into the future when designing an architecture such as the QLA architecture.

---

[1]The movement failure rate is expected to improve from what it is now as trap sizes shrink and electrode surface integrity continues to improve.

7.2   THE LOGICAL QUBIT DESIGN IN THE QLA

The most important component of the QLA architecture is the logical qubit tile, which is built using the recursive methodology we outlined in Section 4.2.4. The low-level geometric structure of the qubit tile is driven by the ion-trap characteristics shown in Table 7.1, which place us significantly below the accuracy threshold value required by the threshold theorem. These parameters are optimistic, but not fundamentally impossible (as suggested by the recent ion-trap literature [114, 130, 156, 193]). Particularly important is the fact that the lifetime of an ion (measured as 14.3 seconds in [130]) is much larger than quantum operations, which are on the order of tens of microseconds. These relatively low memory error rates allow us to significantly reduce the area of a logical qubit by reducing the physical resources necessary for each error correction cycle.

To reduce communication and complexity, each logical qubit in the QLA is modeled as a self-contained hardware structure that requires no external quantum resources to perform logical gates and state stabilization (i.e., error correction). This will allow an application level compiler to divide the quantum program into distinct data independent threads that are executed on separate computational units, which are simply the logical qubits in a homogeneous architecture such as the QLA.

Figure 7.3 shows the full implementation of a Level 2 qubit tile. The two high-level ancilla blocks in a Level 2 qubit allow the error correction of two Level 2 qubits when a two-qubit gate is executed inside a single qubit tile. The two sets of high-level ancilla are necessary in computational tiles to ensure that both logical data qubits are error corrected immediately after the execution of a two qubit gate, without stalling the application execution.

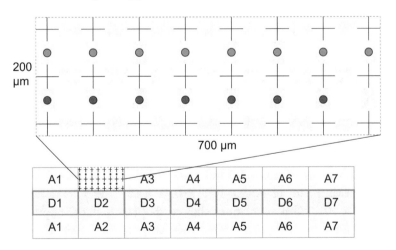

Figure 7.3: The logical qubit tile in the QLA architecture.

As we mentioned in Section 4.2.3, a single data logical qubit at Level 2 is built by encoding 7 Level 1 qubit blocks with the Steane [[7, 1, 3]] code (the [[7, 1, 3]] code allows large set of logical gates to be implemented transversally with the lone exception of the T gate). A Level 1 qubit block is shown at the top of Figure 7.3 A logical CNOT gate is implemented by bringing 49 ions from some qubit tile A in the same trap as the 49 ions in qubit tile B. After 49 CNOT gates are applied on the joined ions, the two sets are error corrected by the ancilla on both sides of the data region in a Level 2 tile. The ancilla preparation network at Level 2 does not require specially designated verification blocks, as the errors are detected during lower-level syndrome extractions [168]. The overhead of the error detection scheme in Reference [168] when preparing Level 2 ancilla is negligible since our device failure probabilities are relatively low.

If physical movement steps are taken into account, the Level 1 error correction circuit shown in Figure 4.7 will take 154 cycles, where each cycle is estimated to be in the order of 10 microseconds and can be as large as 0.003 seconds per error correction procedure at Level 1. In our time estimates, we choose to provide a single laser per Level 1 block. The latency introduced by serializing the Level 1 circuit is not significant since a maximally parallelized circuit would take approximately 127 cycles per error correction procedure. A fully serialized error correction at Level 2 will last approximately 0.3 seconds, which is two orders of magnitude more than the time to error correct at Level 1.

The QLA designers have made the two assumptions when extracting the error syndromes for both Level 1 and Level 2 qubit blocks: (1) two syndromes are extracted in *serial* for both X and Z errors, and (2) it is assumed that in the case of a non-trivial syndrome the next extracted syndrome will match it, and we can proceed with the error correction step. Since the logical qubit at Level 2 is equipped with parallel syndrome extraction, assumption (a) makes Equation 7.1 an overestimate of the final latency:

$$T_{L,ecc} = \begin{cases} 2 \times T_{L,synd}, & \text{Trivial syndrome} \\ 2(2T_{L,synd} + T_1 + T_{L-1,ecc}), & \text{Non-trivial} \end{cases} \tag{7.1}$$

where $T_{L,synd}$ is the time to extract a syndrome at Level L, which is a function of the time to prepare the logical ancilla block. T_1 denotes the time to execute a logical one-qubit gate, and $T_{L-1,ecc}$ is the time for a lower level error correction step that follows each Level L logical gate. A syndrome is considered *trivial* if no data errors are detected; in which case, no error correction is necessary and the syndrome is not repeated to reconfirm the location of any found error. A syndrome is considered *nontrivial* when one or more errors are detected in the data block.

Numerical simulations of a Level 2 qubit showed that a non-trivial syndrome was measured at level one with a rate of $3.35 \times 10^{-4} \pm 0.41 \times 10^{-4}$, and for level two at a rate of $7.92 \times 10^{-4} \pm 0.81 \times 10^{-4}$. Our simulations showed that (assuming the optimistic ion-trap parameters) the syndrome extraction process was always only repeated once. Thus, it is a reasonable assumption that in the case of a non-trivial syndrome, at most, one more syndrome extraction is required before the correcting gate can be applied. Taking a weighted average of the two cases in Equation 7.1, we determine a Level 2 error correction time of approximately 0.3 seconds. As noted in Table 4.1 in Section 4.2.8, using Level 2 recursion with this qubit tile design is sufficient to factor larger than 2048-bit numbers.

We used QASM-TOOLS, to empirically compute p_{th} at Level 2 for the QLA logical qubit. Our results, displayed in Figure 7.4, show that the failure probability of a single one-qubit logical gate rapidly drops to zero at component failure rates lower than $p_{th} = (2.1 \pm 1.8) \times 10^{-3}$. Above this value the rapid decrease in the reliability of our system as recursion increases can be attributed to the additional resource overhead of recursion.

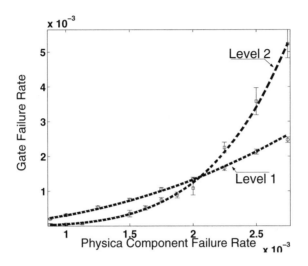

Figure 7.4: Estimate of the failure probability ($\hat{y}$ axis) of a single logical one-qubit gate followed by recursive error correction procedure at levels 1 and 2. The $\hat{x}$ axis denotes individual physical gate failure rates, assuming that they are all equal.

The estimated threshold failure probability is much higher than the theoretical estimate of 7.5×10^{-5} computed in [202] for several reasons: (1) The structure of the qubit is optimized for the error correction circuit and may vary for different codes; (2) The high reliability of ion-trap memory operations has allowed us to significantly reduce the overall area and ancillary resources required; (3) The fixed, low movement error probability, and the fact that we made the design decision to never physically move the data, pushed our qubit's threshold closer to the 9×10^{-3} threshold value estimated by Reichardt [168]. We observed no failure at Level 2 recursion as the physical component errors approached the expected ion-trap parameters from Table 7.1, which was expected. Reevaluating Equation 4.13 with the empirical value for p_{th}, we get an estimated Level 2 reliability approaching the remarkably low value of 10^{-21}.

7.3   LOGICAL QUBIT INTERCONNECT

A logical two-qubit gate between Level 2 qubits $Q1$ and $Q2$ is executed by moving all 49 physical ion-qubits that encode qubit $Q1$ to the computational tile where qubit $Q2$ resides. If the application being executed is the factoring of a 1024-bit number using Shor's factoring algorithm, $Q1$ could be moving as far as 0.5 meters (or 256 logical qubits) across the ion-trap chip. The long-distance communication channel employed by the QLA architecture is the repeater-based teleportation protocol described in Section 4.3, where a repeater station is placed between every logical qubit tile. The ultimate purpose of the repeater-based channel is to create a single EPR pair (we.e. two ions in the maximally entangled state $(|00\rangle + |11\rangle)/\sqrt{2}$) such that one of two qubits is at the location of qubit $Q1$ and the other one at the location of qubit $Q2$. An EPR pair distributed in such a way is required for each of the 49 ion-qubits of qubit $Q1$ (not necessarily created in parallel), such that each of the 49 qubits can be teleported to the computational tile of qubit $Q2$.

Each EPR pair that connects to adjacent repeater stations is created in the middle, where two ion-qubits are entangled and separated to the two opposing ends. There are many ways to achieve entanglement between two ion-qubits. In one scalable entanglement technique for ion-traps [146], the ion-qubits are initialized to the ground state $|00\rangle$ and placed in the same trap. An entangling controlled-phase gate adapted for coupling two ions together is used to implement a CNOT gate (also known as the Mollmer-Sorrensen entangling gate), placing the ions in the intermediate maximally entangled state $(|01\rangle + |10\rangle)/\sqrt{2}$ [191], which can be followed by single-qubit rotations to place the two-ion-qubit state in the desired EPR state.

An alternative proposal [52] combines the features of optical lattices and ion traps, where individual ions are entangled through a common interaction with a pulsed, high-strength optical lattice. The advantage of this proposal is that the two ions do not need to be physically together for the entanglement operation to occur. Using this proposal, however, would drastically change the underlying physical microarchitecture we have described (it is an area worth investigating in the future, however). The Molmer-Sorrensen entangling gate has been used recently in two simultaneous, independent experiments that demonstrate quantum teleportation using trapped ions [18, 171]. To model EPR creation, we assume that two ions are brought together and the actual EPR generation routine is a resource that can be abstracted as a single box (as shown in Figure 7.6) whose implementation can be modeled as the familiar entangling circuit shown in Figure 2.8 from Section 2.8.

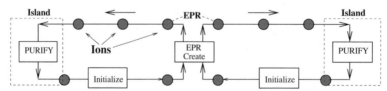

Figure 7.5: Detail of a channel between two repeater stations. The channel is a two-way ballistic transport region where the EPR pairs are created in the middle and distributed in a pipeline fashion to the two Island/Reapeater stations.

To optimize space and performance, we can model the channels between each island as a two-way ballistic transport region as shown in Figure 7.5, which also illustrates the pipeline purification protocol employed by the QLA architecture for purifying a single EPR pair. The basic idea of

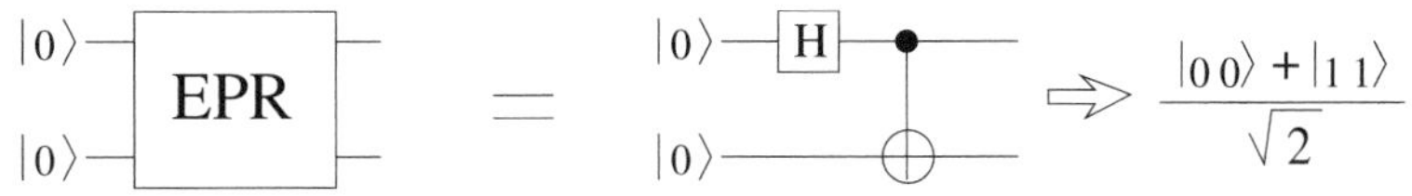

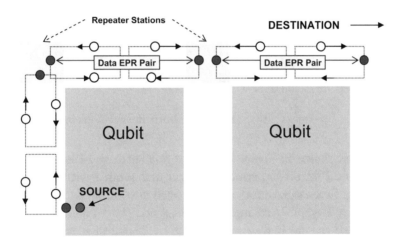

Figure 7.6: EPR Generation can be abstracted as a box or modeled using a Hadamard gate, followed by a CNOT gate between two qubits.

purification [22] is to use several copies of lower fidelity EPR pairs to *distill* a single high fidelity EPR state that can then be used for teleportation. Generally, it is not possible to create a perfect EPR state with unit fidelity, mostly because of the usage of noisy gates in the process of creation and the transmission of the two qubits through the noisy physical channel between each repeater station. In Figure 7.7, a detailed schematic of the pipeline network is shown when repeater stations are placed between every logical qubit. The source ion is being teleported in the direction shown by "DESTINATION" after the channel is prepared.

Figure 7.7: A slightly higher level detail of the communication channel as the repeater stations are placed at the corner of each logical qubit. The ions are not to scale compared to the size of the logical qubit or the pipeline-based interconnect.

If the initial preparation fidelity is high enough, by applying successive purification steps an EPR pair can be purified to an arbitrarily high fidelity. The pipeline purification sequence works by designating one EPR pair as the data pair, which is continually purified in round-robin pipeline fashion by the additional ancillary EPR pairs. An assumption is made that there are enough ion resources in the pipeline to handle the maximum number of required purification steps without having to wait for the creation of new EPR pairs before each successive purification step. The original purification protocol was formulated by Bennett [22], where the efficiency of purification depends highly on the the reliability of the physical gates that make up the protocol (namely Hadamard and CNOT gates) and the initial fidelity of the EPR pair [72]. We use the recursive fidelity equations given in Reference [72] (where the first detailed analysis of quantum repeaters is performed) to study the purification-based repeater protocol employed by our architecture. The efficiency of our design depends as much on the gate reliability as it does on the types of errors that occur and how the errors accumulate in the EPR states before and during purification. This allows distillation higher fidelity EPR pairs with fewer purification steps. The purification circuit is shown in Figure 7.8 where there are two possible network choices. Using the first network in Figure 7.8(a) and limiting purification to be only between two adjacent repeater stations (we.e. islands), we determine sufficient repeater separation to once for every one logical Level 2 qubit.

After its creation (or even during the purification procedure), the data EPR pair accumulates bit-flip or phase-flip errors that can place it in any of the four possible states known as the four *Bell States* $\{|\Psi_+\rangle, |\Psi_-\rangle, |\Phi_+\rangle, |\Phi_-\rangle\}$ [19]:

$$
\begin{aligned}
|\Psi_+\rangle &= \frac{1}{\sqrt{2}}(|00\rangle + |11\rangle) \rightarrow \quad \ldots\ldots\ldots \ \textit{no errors} \\
|\Psi_-\rangle &= \frac{1}{\sqrt{2}}(|00\rangle - |11\rangle) \rightarrow \quad \textit{Z error on q1 or q2} \\
|\Phi_+\rangle &= \frac{1}{\sqrt{2}}(|01\rangle + |10\rangle) \rightarrow \quad \textit{X error on q1 or q2} \\
|\Phi_-\rangle &= \frac{1}{\sqrt{2}}(|01\rangle - |10\rangle) \rightarrow \quad \textit{both X and Z errors}
\end{aligned}
\tag{7.2}
$$

The purification circuit shown in Figure 7.8(a) uses one ancillary EPR pair to check the state $|\Psi_+\rangle = (|00\rangle + |11\rangle)/\sqrt{2}$ for bit-flip errors, and then uses another ancillary EPR pair to check the state for phase-flip (we.e. sign) errors. After interaction with the data EPR pair through the CNOT gates the two ancillary qubits are measured, where odd parity for either X or Z error checks will indicate that there is an error in the data EPR pair. In the case of an error, the data EPR qubits are recycled in the pipeline and the next ancillary EPR pair becomes the data EPR pair that needs to be purified. Each successful purification step increases the likelihood that a data EPR pair is free of errors, thus it increases its fidelity. The principle is the same as flipping a weighted coin with unknown weight, — each time the coin lands heads given that it has landed heads the previous throw, the probability that the coin is weighted towards heads increases.

An alternate purification procedure is shown in Figure 7.8(b), where four EPR pairs are prepared in parallel at the beginning. The data EPR pair is at the top and it is checked in parallel with an additional EPR pair for X errors. If both pass, the data EPR is checked for Z errors. Although we haven't studied this protocol, it may have the potential to offer better purification efficiency by ensuring that the ancillary EPR pair used in the Z error detection is checked against X errors. We found that the probability of X errors slipping through the Z-error purification is negligibly low when the optimistic ion-trap parameters are assumed.

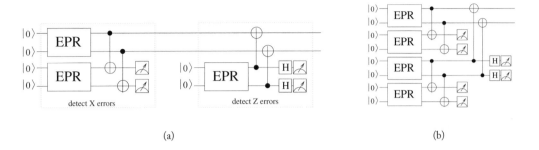

(a)

(b)

Figure 7.8: (a) The data EPR pair (top) is created in parallel with an additional ancillary EPR pair used to detect bit-flip errors first. Phase-flip errors are detected with a third EPR pair or the previous ancillary pair reinitialized. (b) Four EPR pairs are created, two of which are used to check the other two for bit-flip errors. This is followed by the detection of phase-flip errors on the two EPR pairs remaining.

In reality, any of the four Bell states can be used for teleportation, thus the purification efficiency can be further improved if we allow X or Z errors to remain and use the subsequent purification steps to ensure that indeed the X and Z errors detected in the previous step are present. In such cases, we know which of the four Bell states our EPR qubit is in, and to modify the teleportation protocol accordingly where the modification consists of a different interpretation of the 2-bit bitstring that signifies how to apply the correcting X and Z gates on qubit $q3$ in Figure 2.7 at the end of the teleportation protocol.

There are many tradeoffs associated with the microarchitecture design choices made during the construction of the long-distance channels. Suppose we define the *scope* of an EPR pair as the distance between each of the two EPR qubits (as a function of the number of teleportation islands between them). If the entire channel between logical qubits $Q1$ and $Q2$ has K repeater stations, the ultimate goal is to create a single EPR pair with a scope of K islands. EPR pairs that connect two adjacent repeater stations have a scope of zero.

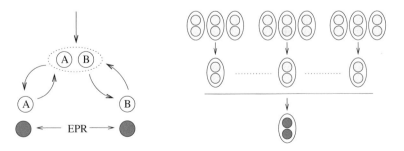

Figure 7.9: Minimum and maximum number of resources needed to purify a single EPR pair.

The first tradeoff when considering microarchitecture design for the communication network is related to the number of ion-qubit resources required to distill a single high fidelity EPR pair spanning any distance. Clearly, the minimum resources required when considering the network in Figure 7.8(a) are four ion-qubits, two for the data EPR pair and two for the ancillary pair used for purification. The ancillary pair is continuously reprepared for each purification step. Alternately, the maximum number of resources necessary can be calculated by creating all EPR pairs required for j purification steps (which would require $\eta(2 \times 3)^j$ ion-qubits, where η is some constant that takes into account the possibility of failure at some stage in the purification).

Both resource extremes are shown in Figure 7.9, where the protocol that uses the minimum resources is shown on the left-hand side. Ion-qubits A and B are continuously reprepared and interacted with the data EPR pair at each step of purification. In the scheme on the right-hand side, a two-step purification tree is shown, where 18 ion-qubits are prepared into three groups of three EPR pairs used for the first purification step. After the first step, three purified EPR pairs are left and used to further distill a single EPR pair. While the first protocol uses far less resources, the final fidelity of the data EPR pair is severely limited by the fact that the ancillary EPR pair is continuously reprepared and retains the same level of noise throughout the purification process. In the second protocol, on the other hand, the data and ancillary EPR pairs are equally purified at each step, and a much higher fidelity is achieved for the final EPR pair. However, this is at the expense of high ion-qubit resources, and it creates a more complex microarchitecture that requires movement of all EPR pairs at each purification step. The pipeline approach, we use as shown in Figure 7.5, allows sequential purification without memory cycle delay between each purification step. By avoiding recursive purification and pipelining the ancillary EPR qubits, we are able to achieve optimal bandwidth requirements (we.e. a bandwidth of a single physical channel) for each distillation of EPR pairs between any two adjacent repeater stations.

A second important tradeoff arises when deciding the separation between two adjacent repeater stations (and thus the distance at which EPR pairs are purified). There are three possible ways to connect a source and a destination separated by a number of repeater islands, such that the final teleportation step of the data qubit between the source and the destination is teleported with the desired threshold fidelity required for error correction:

1. A *purely-linear* approach, which distills high-fidelity EPR pairs only between adjacent islands to some fidelity F that will allow $O(\log K)$ teleportation hops (see Figure 4.15) to be performed such that the final fidelity of the data teleported is within the threshold value. The total time to achieve a given relatively large distance varies as the separation between repeater islands is changed. As the separation decreases, purification will be followed by a greater number of teleportation hops between the source and the destination, requiring more purification to achieve higher EPR starting fidelity. Alternately, as the separation increases, there is a smaller number of teleportation hops, but the data and ancillary EPR pairs travel longer in the pipeline, thus more purification is needed to reduce the fidelity. It is an interesting tradeoff for

a system designer to explore, and it offers an opportunity to design a reconfigurable dynamic interconnect.

2. A *nested, semi-linear* approach, which distills EPR pairs at different nesting levels with an increasing scope per level. This method was analyzed in detail in Reference [72]. At the lowest nesting level EPR pairs are created with a scope of m junctions, which are used to purify an EPR pair with the same scope at the second nesting level. The freshly purified scope m EPR pairs are connected to create an EPR pair with scope km for some other constant k, which are then used to distill a single EPR pair of scope km at the third nesting level. This process is repeated until we have a single EPR pair connecting the source and the destination as shown in Figure 7.10.

3. Finally, we can create EPR pairs directly between the source and the destination without purifying at any intermediate scope. The purification is performed for an EPR pair that spans the source and the destination, until a desired fidelity is reached.

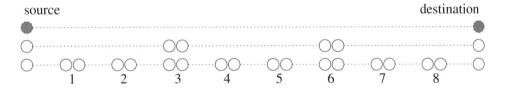

Figure 7.10: Nested purification protocol as described in Reference [72].

The QLA architecture utilizes Approach 1, where we find that at the optimistic technology parameters for ion-traps, the distances required for communication when factoring a 2048-bit number (maximum across 512 logical qubits) are attainable when the separation between two adjacent repeater islands is 5 logical qubits.

Approach 3 was studied in detail in Reference [101], where the creation of hundreds of EPR pairs is required between the source and the destination to purify EPR pairs that span the entire channel. Intuitively, both approaches, 2 and 3, are able to achieve much longer communication channels between two distant logical qubits than the linear approach employed by the QLA architecture. For example, when the movement failure rate is reduced by an order of magnitude to $O(10^{-5})$, the QLA interconnect can send qubits across only 10 logical qubits, while Approach 3 allows qubits to be sent across as much as 221 logical qubits and still attain final fidelity within the threshold fidelity for computation. This approach, however, is very expensive when one considers the necessary EPR resources. Our simulations indicate that 4 purification steps are required to bring the fidelity to the threshold fidelity of $1 - 7.5 \times 10^{-5}$. This means that 16 EPR pairs must be sent across the entire channel to distill a single good EPR pair (actually 32 must be sent to purify for both bit-flip and phase-flip errors). In addition, this does not take into consideration the additional EPR pairs required within each two adjacent repeater stations. Greater distance can be achieved in both nested repeater protocols, at each purification step the fidelity of the data EPR pair and the fidelity of the ancillary EPR pair improve equivalently, while in the purely linear approach employed by the QLA, the fidelity of the ancillary EPR pairs remains fixed (a function of the separation distance between adjacent repeater stations).

The QLA's pipelined linear approach offers a comparatively smaller bandwidth by providing only a single pipeline based channel from the source to the destination. The tradeoff is the serialization of the purification process, however, because the purification is pipelined, the only temporal cost suffered is an initial purification cost which increases linearly with the separation of the repeater stations. The initial cost is the time of creating the first ancillary EPR pair and the time it takes to transport the pair to the two adjacent repeater stations. Estimates of the interconnect performance are given in Figure 7.11. On the left-hand side in Figure 7.11(a), the total connection time for teleporting 49 physical ions sequentially is plotted as a function of the desired communication distance. We see that a repeater separation of one logical qubit is sufficient to allow us to communicate all 49 physical ions sequentially across a distance sufficient for applications as large as factoring a 1024-bit number.

One important observation in the plot of Figure 7.11(a) is that the total connection time is dominated by the number of purification steps and thus, by the separation of the repeater stations. For a total distance of just below 300 qubits, the total communication time behaves logarithmically as expected, but jumps suddenly each time additional purification steps are required. For the factoring of a 1024-bit number with maximum communication distance across 256 logical qubits and repeater separation of 1 logical qubit, we need at most 3 purification steps.

The plot shown in Figure 7.11(b) shows how the total communication time increases as the separation between repeater islands increases. After a maximum distance peak, the total possible distance decreases as the separation increases, simply because the EPR qubits cannot be purified enough. The increase in total connection time is linear with the increase in the island separation.

In quantum architectures the cost of communication is just as critical for application execution as the cost of logic. We have made a design decision that ballistic transport be used for moving ions within a logical qubit, and teleportation will be preferred when moving across larger distances (in order to keep the failure rate due to movement below the threshold amount). Since EPR pairs are required for teleportation, we can reduce communication costs to a minimum if we have the required number of EPR pairs available at a logical qubit at the same time that it is ready to move. This overlap of computation and communication is possible because of the high latency of each error correction step.

Our results (shown in Figure 7.11) indicate that we can create, purify and transport the required EPR pairs to their respective qubits while they are undergoing error correction. But can this be done at a large scale? In other words, given a mapping of the high-level circuitry for some quantum subroutine such as the quantum Fourier transform, the central question is if there are enough logical interconnect channels free that will allow us to pre-prepare them for teleportation (we.e. distribute the needed spanning EPR pairs) while the source qubit is undergoing error correction.

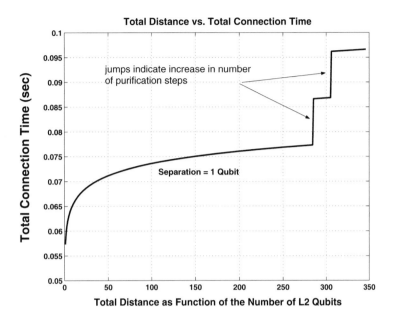

(a)

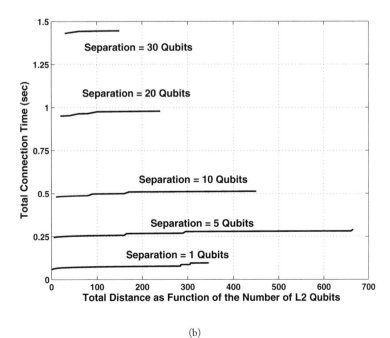

(b)

Figure 7.11: Caption on the next page.

Figure 7.11: (a) Total time to teleport 49 qubits sequentially as a function of the total communication distance for repeater separation of 1 logical qubit. (b) Same as (a), but with varying repeater separation. In each figure, we performed enough purification steps such that the failure rate of the teleported data qubit is below the threshold estimate of 7.5×10^{-5}.

To answer this question, we used a simulation tool to schedule the movement of EPR pairs in QLA [140]. We assigned one channel to carry the created EPR pairs to their destinations and another channel to return the used EPR pairs. Within each channel, the EPR pairs are pipelined. We define the bandwidth of QLA's communication channels as the number of physical channels in each direction - the channel shown in Figure 7.5 has a bandwidth of 2. The goal of the scheduler is to find paths between logical qubits to transport all the required EPR pairs within the time it takes to perform a Level 2 error correction.

	N=128	N=512	N=1024	N=2048
Logical Qubits	37,971	150,771	301,251	602,259
Toffoli Gates	63,729	397,910	964,919	2,301,767
Total Gates	115,033	1,016,295	3,270,582	11,148,214
Area(m^2)	0.11	0.45	0.90	1.80
Time(days)	0.9	5.5	13.4	32.1

The scheduler is a heuristic, greedy scheduler that works by grabbing all available bandwidth whenever it can. However, if this means that the scheduler cannot find the necessary paths, it will back off and retry with a different set of start and end points. A simple approach to doing a two qubit gate between logical qubits Q1 and Q2 would be as follows: teleport Q1 to Q2's physical location, perform the gate and teleport the result back. An optimization that the scheduler incorporates is that it only moves logical qubit Q1 back if necessary. As a result, the logical qubits *drift* from one location to another. This adds a level of complexity to the scheduler, but at the same time reduces the amount of movement that the qubits are subjected to. With all of the above considerations in the scheduler, we found that given a single pipeline for each channel (as shown in Figure 7.7), we can schedule communication such that it always overlaps with error correction of the logical qubits. The end result of the communication computation overlap is that logical qubits can cover sufficiently large distances with minimal overhead on the application execution time.

To test that computation and communication can indeed be overlapped and to estimate the performance of the architecture, we used the above methodology to schedule onto the QLA the high-level execution for both quantum subroutines of Shor's factoring algorithm: the modular exponentiation routine and the quantum Fourier transform. The modular exponentiation routine divides into a series of modular multiplications. Each modular multiplication divides into a series of quantum adders. Each subroutine is modeled using the universal gate set given in Equation 2.10.

The results for the estimated performance of the homogeneous QLA architecture for Shor's algorithm when considering its two quantum subroutines are summarized in Table 7.2. Area numbers assume current experimental trap sizes of about 20 μm traps, which is also assumed to be the size of each cell in the QLA layout. The numbers in the table are for a single execution of modular exponentiation and do not include the expected number of times the whole calculation will need to be repeated. The large factor between N and the number of logical qubits and gates comes from the fact that many logical qubits are needed to implement the Shor circuit (in addition to the core data qubits) and due to the fact that error correction adds to the overal qubit overhead.

7.4   COMPRESSED QLA ARCHITECTURE: CQLA

The design of the QLA architecture has so far followed the conventional wisdom of maximizing parallelism to improve both the run time of the application and to decrease the active time of the qubits. To achieve maximum parallelism, we have abstracted each logical qubit tile as its own computational unit where encoded program operations can occur. To improve the error correction performance and reduce the accumulation of errors, each logical qubit tile has a (data:ancilla) ratio of (1 : 2) between physical ions used to store encoded logical data and physical ions used to store encoded high-level ancilla for error correction.

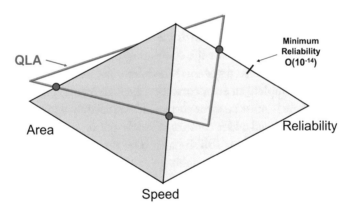

Figure 7.12: Homogeneous QLA representation on the design pyramid, where the area is not good, the performance is acceptable and the achieved reliability is excellent given the application requirements.

The area consumption of such a design is estimated to be about one square meter to factor a 1024-bit number [140], which can be untenably large. Given that there are no large-scale architecture models to compare the QLA to, we represent the homogeneous QLA on the design pyramid as shown in Figure 7.12, where the area is shown to be problematic. the performance is acceptable (considering 10 μs clockcycles), and the operation reliability very good, $O(10^{-21})$ (given the optimistic ion trap parameters). As shown in the pyramid, the homogeneous QLA is a very unbalanced system design.

To make the CQLA architecture more balanced, we investigated if it is possible to trade some of the excess reliability and improve both the area and the performance.

In Chapter 6, we discussed performing computation and storage in separately constructed logical qubit tiles. We even speculated that it may be beneficial to encode data differently between compute tiles and memory tiles, a design choice which may help us reduce the area introduced by the homogeneous architecture, and hopefully improve the time performance of the computer. However, the work presented in Chapter 6 showed that the simplest way to reduce the area requirement is to leave the level of recursion and error correcting codes unchanged, but designate some qubit tiles for computation and some for data storage. Counterintuitive to classical architectures, the tiles that allow encoded gates to be applied on the data contain more error correcting resource to allow faster error correction operation after each logical gate.

Higher physical ion density, in terms of ion-qubits that store data per unit area, can be achieved in the memory region by increasing the ratio between physical ions that store data and physical ions used to correct the encoded data (as shown in Figure 6.2 in Chapter 6). By surrounding a single logical ancilla block by eight logical data blocks to form one memory tile, we can increase the error correction cycle time per logical data block, but we *increase* (data:ancilla) ratio from (1 : 2) to (8 : 1). Even with the increased ratio, memory error rates remain much better than gate error rates ad the probability that an ion-qubit will fail while waiting for the next error correction cycle is within the accuracy threshold value of the $[[7, 1, 3]]$ code. Additionally, when a logical data qubit residing in a memory tile is needed for gate execution while waiting for the next error correction procedure, the teleportation of the logical qubit combined with the error the data has accumulated while waiting may introduce too many physical errors for the computing tile to recover the logical qubit once the data is teleported there. In such a case, a systems scheduler must issue a "stall" for the operation and wait for the logical data to complete an error correction cycle in memory before it can be teleported to the compute region, where it must be error corrected immediately upon arrival.

The Draper carry-lookahead adder consists of single-qubit gates, two-qubit CNOT gates and the gate count is heavily dominated by Toffoli gates. The time to perform a single fault-tolerant Toffoli is equal to the time for fifteen two qubit gates (including T gates, which are essentially two-qubit gates), each of which is followed by an error-correction step. Table 7.4 shows the area savings that can be achieved when using denser memory for various adder sizes. The table shows that the performance is slightly negatively impacted for the $[[7, 1, 3]]$ error correcting code. To limit the performance degradation, we have addressed the parallelism available within the application itself and determined the number of compute blocks necessary to maximally exploit this parallelism with changes in the problem size N (we.e. factoring an N-bit number). For a fixed problem size, utilization of each compute block decreases with an increase in the number of compute blocks as shown in Figure 4.13(b). Clearly, the decrease in utilization is offset by the increase in overall performance. Thus, the challenge in this case is to find the *balance* between utilization and performance.

The new design pyramid is shown in Figure 7.13, where we have improved the area by reducing the ancilla to data ratio, at the expense of some of the excess reliability and some of the speed. The

Input Size	Compute Blocks	Area Reduced (Factor of)	SpeedUp	Gain Product
32-bit	4	6.69	0.54	3.61
	9	3.22	0.97	3.14
64-bit	9	6.36	0.70	4.45
	16	3.79	0.98	3.71
128-bit	16	7.24	0.72	5.24
	25	4.90	0.96	4.70
256-bit	36	6.65	0.92	6.12
	49	5.07	0.98	4.96
512-bit	64	7.42	0.92	6.80
	81	6.06	0.98	5.94
1024-bit	100	9.14	0.80	2.19
	121	7.81	0.97	2.65

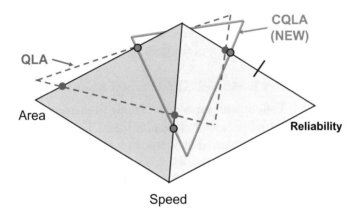

Figure 7.13: The first optimization of the design pyramid. By separately optimizing the ancillary resources for computation (cycle speed) and storage (data per unit area), we have improved the area considerably, but have lost a bit of performance and reliability.

reliability is negatively impacted because we increased the time a data qubit spends in memory between error correction cycles, thus allowing more memory errors to accumulate. Similarly, the overall performance is degraded due to limiting the parallelism and increasing the memory cycle time, thus introducing some additional latency between operations. Before we address (in Section 7.4.3) the performance degradation by trading even more reliability, we will discuss first a new architecture performance metric we have introduced to mitigate the different tradeoffs associated, *the Gain Product*, and some details about the execution of the Toffoli gate.

7.4.1   THE GAIN PRODUCT: ARCHITECTURE PERFORMANCE COMPARISON

When the overall performance of the new specialized architecture is compared to the homogeneous QLA architecture [140], we see that area is reduced by a factor of 9 when factoring a 1024-bit number using only 100 compute blocks. The performance reduction is almost 20%, where the underlying error correcting code remains the $[\![7, 1, 3]\!]$ code. Since one of the most feasible large-scale ion-trap schemes requires the electrodes to be etched into a Silicon substrate [38, 112], equal performance must be placed on the system area and the execution speed.

A good metric we introduce for evaluating the merit of our design choices that affect both area and computational speed is the *Gain Product* (GP) [60], defined as:

$$GP = \frac{(Area_{old} \times ExecutionTime_{old})}{(Area_{new} \times ExecutionTime_{new})},\tag{7.3}$$

where *ExecutionTime* is the execution time per application procedure. In this example, *ExecutionTime* is defined as the average time per adder for modular exponentiation. The gain product (GP) indicates the improvement in system parameters normalized to the homogeneous QLA architecture, which is assumed to have a GP value of 1. The higher the gain product, the better the collective improvement in area and time of the system.

7.4.2   COMMUNICATION ISSUES: EXECUTING THE TOFFOLI GATE

As described in Section 2.4, Toffoli gates cannot be directly implemented on encoded data and must be broken down into multiple one and two-qubit gates. The one-qubit gates include the T gate, which requires the specially prepared encoded $A_{\pi/8}$ ancillary state to execute the T gate implementation shown in Figure 4.9 in Section 4.2.3. The circuit used for preparation of the $A_{\pi/8}$ ancilla state is described in detail in Reference [7] and is given in Figure 7.14, where we see that the preparation requires two error correction steps and measurement of an additional two 7-qubit cat-states. Separate *specialized ancilla* tiles must be designated to ensure a fresh supply of prepared $A_{\pi/8}$ ancilla states be available when needed. The execution of a T gate requires both accumulators be in a single computational tile, where one accumulator is occupied by the data qubit and the other by the $A_{\pi/8}$ ancilla.

The flow of data between the three qubits to complete a single Toffoli forms the most intense communication pattern during the entire addition operation. To study the bandwidth requirements needed by the Toffoli gates, a heuristic greedy scheduler was created that attempts to have the requirements for communication (creating EPR pairs, transporting and purifying them) in place while the logical qubit to be transported is undergoing error correction after completion of the previous gate. As it turns out, with the bandwidth of a single channel as shown in Figure 7.5, it may be possible to completely overlap communication and computation when using the the Steane $[\![7, 1, 3]\!]$ code.

Superblocks: The notion of a *superblock* is defined to mean a collection of one or more computational tiles. The computational tiles are grouped together to exploit the principle of *locality* inherent in

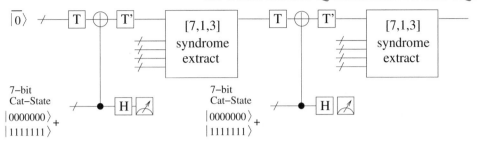

Figure 7.14: The circuit for the creation of a single $A_{\pi/8}$ ancillary state.

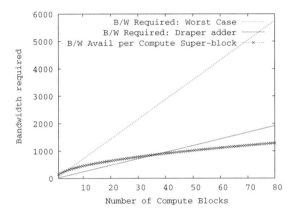

Figure 7.15: The point of intersection of the two bottom curves is the optimal size of a compute superblock. These two curves are bandwidth required (at the perimeter of the compute superblock) in modular exponentiation and bandwidth available. The third steep curve is the worst case bandwidth required.

quantum applications, which (much like the classical definition) means that data is most like to be reused soon after each usage. Larger superblock regions have the advantage of an increased perimeter bandwidth between the compute and memory regions of the specialized architecture, which is offset by the much greater increase in communication required by having to move data from the computing region to the memory region. Intuition suggests that at a certain point, it may be more efficient to have multiple small superblocks instead of one large superblock. The authors in Reference [60] explore this notion and determine this number concretely. In Figure 7.15, we show the change in bandwidth required against change in bandwidth available as the number of compute blocks increases. The cross-over point is 36 compute blocks per superblock, immaterial of what error correction code is used. Thereafter it is no longer beneficial to increase the size of an individual compute superblock.

7.4.3   MEMORY HIERARCHY IN THE CQLA ARCHITECTURE

In our discussion on further optimizing the QLA architecture such that the performance is improved as well, we finally discuss the notion of the quantum cache. In fact, the QLA discussed in the previous section does not even have a memory hierarchy to consider, for the computational tiles and the memory tiles were both constructed using the same error correcting code at the same level of recursion. We saw how the mere separation between memory and computation (i.e., when decoherence of qubits in memory is smaller than qubits in computation) can dramatically reduce the area of the quantum processor. But what about performance? The most straightforward way to impact performance is to consider the level of encoding in the computational tiles. Consider, for example, if some of the Level 2 encoded qubit tiles are actually Level 1 encoded tiles. If this were the case, there are two problems that are being introduced:

- If the Level 1 tiles are computational tiles, then the Level 2 memory latency significantly impacts the application performance due to the slower memory cycles. A Level 1 computational tile will simply not have the error correction efficiency necessary to overcome the memory error accumulation during a Level 2 memory read or write cycle.

- Perhaps even more importantly, the reliability of the logical operations from Level 1 to Level 2 improves *doubly exponentially*, and thus for large-scale applications it may be necessary that everything really is at Level 2 encoding.

However, the benefits of Level 1 versus Level 2 encoding are significant. Logical gates are much faster at Level 1 than they are at Level 2, and the amount of reduced area when using 7 instead of 49 ion-qubits per logical data qubit is significant. The cache makes this encoding scheme work by utilizing the principle of locality and by being a buffer between the encoding in the processing elements and the encoding in memory. The data that resides in the cache is placed there either from the processing elements, or has been teleported there through the transfer network shown in Figure 6.6 in Chapter 6. Recall that the transfer network is a tile-based computational structure that implements the process of code teleportation to prevent decoding the data between its transfer from one encoding to another. Once again, the cache is at Level 1, and we must account for the loss in reliability of the logical data stored there.

To see why the loss in reliability cannot be disregarded, consider that the failure per component for the entire system of size $S = K Q$ must be at most $1/K Q$. Suppose that all operations required by a given quantum application and performed by the QLA architecture are divided between Level 1 and Level 2 operations. A crucial observation here is that error correction cycles on a logical data residing in the main memory is considered a logical operation on the data, with an associated level of reliability and an action that does not change the state of the qubit. Even error correction is performed using noisy lower-level gates and can introduce errors on the data. The QLA architecture now consists of a memory at Level 2, a compute region also at Level 2, a cache and a compute region at Level 1 and transfer networks for changing the qubit encoding levels. A revised estimate of the required failure per component is needed to account for the loss of reliability due to the Level 1 encoding.

An intuitive interpretation of the KQ system size parameter is that it is the area of a 2-D rectangle, where one side is the number of logical qubits and the other side is the length of the computation as a function of the number of timesteps K. The area of the rectangle can be divided into several regions: (1) The region of operations that take place at Level 1; (2) The region of operations that take place at Level 2; (3) The region where qubits are "dead" (whose states have not yet been initialized); and (4) The transfer regions between Levels 1 and 2. The transfer region is actually divided into logical operations between Levels 1 and 2, so there is no need to distinguish between operations performed during the transfer and operations performed in the computational region (we just need to consider the added overall operations for the application). In addition, the third region of "dead" qubits is insignificant for the overall system KQ value while executing an application like Shor's algorithm, because after only the first few timesteps all qubits have been initialized and used by the evenly distributed adders. With the different level of encoding, the modified desired crash failure probability per component is lower-bounded by:

$$\epsilon_{fail} = \frac{1}{f_{L1}KQ_{L1} + f_{L2}KQ_{L2}},\qquad(7.4)$$

Where f_{L1} is the fraction of the time spent computing at Level 1 and f_{L2} is the fraction of the time spent computing at Level 2. The total KQ parameters for pure systems at Level 1 and Level 2 are denoted by KQ_{L1} and KQ_{L2}, respectively. The crash failure rate at Level 2 is much smaller than the one at Level 1, so we cannot divide the total operations evenly between the two regions. It is also incorrect to assume that the longer the data stays at Level 2, the more reliable it becomes, and the more operations we can have at Level 1 before a failure occurs. The moment the encoding of a logical qubit is teleported to a different level of error correction, the first error correction cycle or logical operation must ensure that the qubit does not accumulate more errors than the new level of encoding can handle. The reliability of the encoded qubit immediately becomes controlled by the new encoding, and the time we can compute or store the qubit at that encoding state is controlled by the KQ parameter of the subprocedure executed with the data in question.

We find that, to sustain scalability for Shor's algorithm in the QLA architecture, we can perform at most only 1% of the total operations at Level 1, while the rest must be at Level 2. One can imagine this to mean that we can only spend 1% of the total logical cycles over all qubits at Level 1 recursion, including error correction cycles. Specifically, for Shor's algorithm, however, we find a surprising result: Since quantum modular exponentiation is performed by repeated quantum additions, we find that to comfortably maintain the fidelity of the system, we can perform one Level 1 addition for every two Level 2 additions. All the operations performed at Level 1 this way constitute only 1% of the total system KQ parameter, should we have performed everything at Level 2. The resulting increase in performance is shown in Table 7.4. A key note here is that, over the entire system KQ rectangle, the adders implemented with Level 1 logical operations constitute less than 1% of all time cycles.

Par Xfer	Adder Size	L1 SpeedUp	L2 SpeedUp	Adder SpeedUp	Area Reduced	Gain Product
Steane [[7, 1, 3]] Code						
10	256	17.4	1.0	6.2	5.1	31.7
	512	17.4	1.0	6.3	6.1	38.4
	1024	18.2	0.9	5.0	9.1	45.1
5	256	10.4	1.0	4.1	5.1	25.0
	512	10.4	1.0	4.0	6.1	24.5
	1024	11.0	0.9	2.9	9.1	26.9

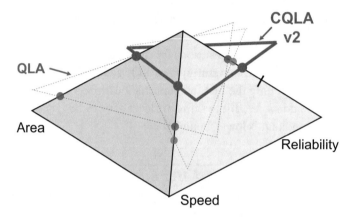

Figure 7.16: The final balanced architecture design when reliability has been further traded by lowering the level of recursion to improve the performance.

The design pyramid that illustrates the final balanced architecture model is shown in Figure 7.16. We have traded additional reliability by lowering the level of recursion for some of the computational timesteps, but have drastically improved the performance.

7.4.4   SIMULATING THE CACHE IN THE CQLA

The only communication between the QLA architecture and the classical control processors is the results of measurement and commands for executing classically scheduled quantum instructions. All communication patterns and instruction order execution are orchestrated through software tools that run in the classical processors. To study the behavior of a specialized architecture into software-managed caches or scratchpad memory we have created a simulator that models a cache as described in this section. The simulator takes into account the computation cost in both encoding levels and also the cost of transferring logical qubits between encoding levels. The application under consideration is still the Draper carry-lookahead adder [69]. Input to the simulator is a sequence of instructions where each instruction is similar to assembly language for quantum computation and describes a logical gate between a number of qubits.

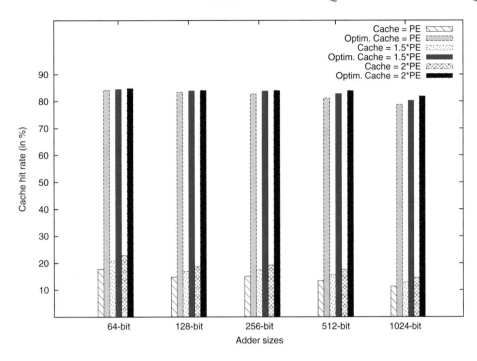

Figure 7.17: The cache hitrate for different adders when both cache and compute region are at Level 1 recursion.

When the simulator runs this code in the sequence intended by the Draper carry-lookahead adder, the cache hit-rate is limited to 20%. To improve the hit-rate, the authors in Reference [60] utilize the following optimized approach. Since the scheduling is static (i.e., run-time instruction scheduling is not assumed at this stage of development), the instruction fetch window for the simulator can be the entire program being executed. The simulator takes advantage of this by first creating a dependency list of all input instructions. Then it carefully selects the next instruction such that probability of finding all required operands in the cache is maximized. This *optimized fetch* yields a cache hit-rate of almost 85%, regardless of adder and cache size. The replacement policy in the cache is *least-recently-used*. Figure 7.17 shows the cache hit-rates for different sized adders for the non-optimized and optimized instruction fetch approaches. If n is the number of logical qubits in the compute region, the cache sizes studied are n, $1.5n$ and $2n$. As the graph shows, the increase in hit-rate is more pronounced due to the optimized fetch than due to increasing cache size. A sensible choice for the QLA architecture, then, is to use a cache whose size is twice the number of qubits in the compute region. The high hit-rate means the complex transfer network of Figure 6.6 will not be overwhelmed.

7.5   QUALYPSO

We've seen how the CQLA architecture improves upon the QLA architecture by providing us with different types of tiles designed for compute and memory operations. Of course, it's possible to envision an architecture where we allow even more types of tiles (each with their own memory, compute, and error correction capabilities). The Qualypso architecture (introduced by Isailovic et al. [103]) is an example of an architecture with such a diversity of tiles.

Figure 7.18(a) shows an overview of the Qualypso architecture. It contains several kinds of compute and memory tiles, each with different ancilla generation capabilities. Each tile (as shown in Figure 7.18(b)) consists of a data region surrounded by an ancilla 'factories' that prepare ancillae for use within the tile and a set of output ports. Qubits are moved ballistically within a tile, so unnecessary teleportation operations (which may require additional error correction) can be avoided if large data tiles are used.

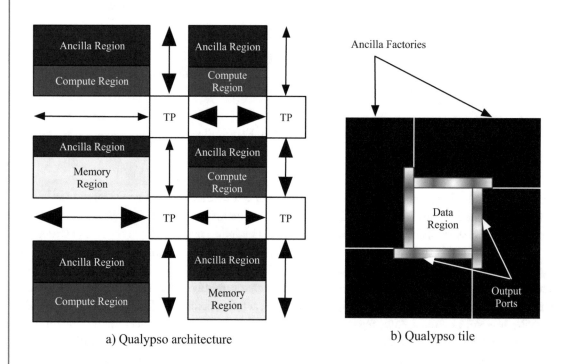

a) Qualypso architecture   b) Qualypso tile

Figure 7.18: a) Qualypso architecture. b) Single tile of Qualypso architecture.

The size and structure of the each of the tiles can vary depending on the ancilla generation requirements of the application the architecture is being adapted for. In [103], Isailovic et al. provide a set of numerical results on the the ancilla preparation latencies and requirements for various circuits. These include a 32-qubit quantum ripple-carry adder from [68], a 32-qubit quantum carry lookahead adder from [70], and a 32-qubit QFT circuit. Table 7.5 shows a set of latency estimates (built using physical parameters from experimental work on trapped ions) for data operations, error correction operations involving data qubits, and ancilla preparation operations.

Circuit	Data Op. Latency (μs) (% of total)	Data QEC Interact Latency (μs) (% of total)	Ancilla Prep. Latency (μs) (% of total)
32-Bit QRCA	29508 (5.2 %)	95641 (16.7 %)	447726 (78.2 %)
32-Bit QCLA	3827 (5.3 %)	11921 (16.7 %)	55806 (78.0 %)
32-Bit QFT	77057 (5.0 %)	365792 (23.7 %)	1097376 (71.2 %)

As indicated by the last column in Table 7.5, ancilla preparation times dominate the total latency of each circuit. For this reason, it's beneficial to dedicate a sizeable portion of tile resources to generating a steady stream of aniclla.

One notable technique for generating a stream of ancilla qubits is to utilize a pipelined aniclla generator. Designs for such generators are also investigated by Isalovic et al. in [103] in the context of attempting to improve the area utilization of the design. Figure 7.19 shows the stages of a pipelined ancilla factory (also from [103]) for producing encoded zero states. It consists of four stages: zero preparation, controlled X-operations/Cat state preparation, verification, and bit and phase correction. Note that qubits from the verification and correction stages are reused at the beginning of the circuit where stateless qubits are zeroed.

Evaluation of this pipelined design produces uncovers mixed results. Compared to a non-pipelined approach, the pipelined approach delivers almost 3.4 times as much throughput but requires approximately 3.1 times as many macroblocks. Hence, the pipelined approach does not produce a drastic improvement in area utilization. However, an application that requires a single high-bandwidth source of ancillae can benefit the significantly improved throughput of the pipelined ancillae generator.

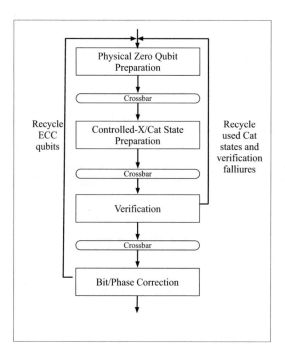

Figure 7.19: Pipelined ancilla generator from [103]

CHAPTER 8

Programming the Quantum Architecture

In this chapter, we discuss how a quantum architecture (such as the QLA) can be programmed. The programming model for the QLA architecture is guided by the fact that each logical qubit tile can be regarded as a separate processing element, designed to execute a *localized* piece of the larger application: a single logical gate followed by error correction. QLA becomes an architecture composed of reconfigurable computational units, where each unit is a self-contained logical qubit tile. By decoupling communication and computation, the QLA becomes an architecture composed of reconfigurable computational units, where each unit is a self-contained logical qubit tile. Creating the timed physical pulse sequences for a quantum application subroutine, therefore, follows the classical GALS model: Globally Asynchronous (at the application level) and Locally Synchronous (within each logical qubit tile the error correction procedures reuse the same pulse-sequence of operations).

This chapter begins with a description of the Instruction Set Architecture (ISA) which is the interface between a quantum compiler and the quantum hardware. At the application level, we have logical instructions acting on logical qubits, where measurement operations give the control processors knowledge about the algorithm execution. Below the application level is the physical layout, where basic logical gates are decomposed into a fault-tolerant sequence of elementary, technology-dependent, assembly-like instructions. The discussion in this chapter is focused on the description of the high-level instruction set architecture, which is independent of the physical implementation technology and allows the compiler to orchestrate the architectural resources available.

The machine instructions that will be described operate on both quantum data (logical qubits) and classical data (such as logical qubit addresses, measurement results and classical control bits). All classical data is stored and manipulated by the classical control processors. The only access the classical processors have to the quantum hardware is through the execution of measurement instructions, which contain both classical and quantum arguments.

A summary of some of the suggested instruction types is shown in Table 8.1. Most instructions at the architecture level of a quantum computer can be classified as *procedure call* instructions. For example, a basic quantum gate instruction could be "gate_cnot Q1,Q2", where Q1 is the control logical qubit and Q2 is the target. If the gates are logical gates, then they are implemented in the hardware using a fault-tolerant sequence of operations on lower-level qubits. Therefore, each logical gate is a self-contained computer program in itself that is incorporated into the larger application and is separately optimized. The control processors are instructed to execute the entire group of lower-

Instruction	Arguments	Type	Function
GATE_A	Q1, [Q2]	Quantum	Single or Two-Qubit Operations
MEASURE	Q1, cbit	Quantum	Measurement, result stored in the classical cbit
PREPARE	Q1	Quantum	Prepare an initial qubit state
FETCH	Q1, PE_i, AC_j	Classical	Fetch qubit Q1 into PE_i and accumulator j
SEND	Q1, memory_address	Classical	Send Q1 from a PE into memory (cache or main)
LOAD/STORE	Q1, memory_address	Classical	Load/Store Q1 to the specified address
REFRESH	Q1	Classical	Error correct qubit Q1

level operations necessary for the completion of a logical CNOTgate. Error correction procedures are also implemented as a single instruction, with a single logical qubit as an argument. Branching instructions serve the same purpose as classical branches, though the decision of whether to branch or not is always dictated by a classical bit set by the result of a quantum measurement.

A similar ISA is given by the low-level quantum assembly language (QASM), first proposed and implemented by Balensiefer et al [13, 14]. QASM consists of a sequence of declarations and commands for physical qubits, similar to the logical procedures shown in Table 8.1. Qubits, classical bits, gate names, and classical functions are initially declared. The preparation procedure indicated using the two physical gates XPREPARE and ZPREPARE, which place a qubit in either eigenstate of the X or Z operator, respectively, and can be decomposed into a measurement operation followed by a corresponding single-qubit gate. The ZPREPARE operation is implemented by applying a measurement gate, followed by a bit-flip gate if the measurement result yields a "1". The qubit is this way initialized to the $|0\rangle$ state. The XPREPARE gate places the qubit in either the $|+\rangle$ or $|-\rangle$ state by applying a Hadamard operation on a qubit prepared in the |0> state.

8.1   PHYSICAL-LEVEL INSTRUCTION SCHEDULING

Irrespective of the underlying technology used to implement a circuit-model based quantum system, one of the central challenges of accurately modeling the physical components of a large-scale architecture is the ability to map and schedule a quantum application onto a physical layout by taking into account the cost of communication, the classical resources, and the maximum parallelism that can be exploited. Physical schedulers based on traditional classical scheduling techniques may be useful in this area. Just as in classical superblock schedulers, the output of a quantum physical operations scheduler can also be a QASM file, but one that is fully parallelized and into which communication instructions have been inserted.

In addition to generating two-dimensional information about the communication paths for a given quantum circuit, a physical operations scheduler allows the determination of the exploitable instruction-level parallelism (ILP) in quantum circuits. Studying the limits of ILP can be used by hardware designers to avoid spending resources on classical control features that will remain underutilized throughout the computation. Furthermore, the existence of massive ILP is an underlying requirement for achieving the optimum possible schedule in quantum error correction [4]. Even though it has been shown that a threshold value exists when movement is considered [86, 201, 202], the ability to *precisely* predict the amount of communication during error correction is crucial for determining how high the threshold value really can be. In addition, knowledge of the communication requirements and available ILP will provide us with better understanding of the exact hardware resources needed for error correction.

The QUALE tool-chain from the University of Washington [14] uses the classical Path-Finder package [139] to map an instruction in a quantum circuit onto a physical layout, provided that it is known ahead of time which qubits are supposed to move. Alternatively, we have developed the physical scheduler QPOS, which is a quantum physical operations scheduler based on traditional classical instruction scheduling heuristics [46, 49, 62] through careful priority calculation at both the circuit level and the physical layout level, without assuming a particular technology implementation. QPOS is described in detail in Reference [141] and is designed to generate the exact communication pattern for quantum data given a specific geometrical layout. At the circuit level, instruction priorities are based on the number of instructions that depend on each operation. The priorities are used to choose the desired communication paths after the source qubits and the destination qubits have been disambiguated. If instructions have the same priorities the paths are prioritized based on least path interference and shortest path. This attempts to maximally parallelize the movement operations.

QPOS uses the generic layout abstraction for quantum devices at the hardware design level as shown in Figure 7.3 of Chapter 7. The physical ions are placed on any of the lines known as interaction regions, and ions are shuttled through the shared junction resources marked by the regions where the lines cross. Data input for two-qubit gates requires the placement of two qubits in the same interaction region. No gates are allowed in any junction, however, the specification of the junctions is necessary to avoid the overlap of qubit data-paths. The scheduler is a heuristic which accepts an application description as a quantum assembly language file, together with a physical layout description, and outputs a sequence of operations that include the available Instruction Level Parallelism (ILP) that can be extracted given the communication paths of each qubit.

While recent breakthroughs in error correction algorithms [7, 11, 164] combined with clever large-scale quantum architecture design (such as the specialization of the QLA architecture) allow us to be optimistic about the realization of applications such as Shor's quantum factoring algorithm [184], the precise orchestration of millions of interacting physical qubits at the hardware level, will undoubtedly prove to be necessary for the design and optimization of realistic device implementations. For example, one of the central challenges that remains for the realization of a large-scale ion trap quantum computer is the modeling and engineering of the interfaces between the large number of physical ions (qubits) in the computer and the classical apparatus that controls them [114]. In Section 8.5, we describe how we can use QPOS to aid in the design of efficient ion-trap systems in order to speed up their development and presever the threshold value necessary to support fault-tolerant quantum computation.

8.2 HIGH-LEVEL COMPILER DESIGN

We can identify several levels of "reordering rules" that a quantum compiler may employ at any stage of the compilation. On one level are circuit optimizations independent of the underlying architecture (i.e., communication cost is not considered, and each gate has a unit cost). The next level of algorithm optimization/compilation is the *coupling* of computation and communication, where the cost of a given high-level considered when the circuit is mapped to a specific set of logical hardware resources. A third level of optimization is when error correction and communication methods at the logical level of application execution are used to aid in the synthesis of logical quantum circuits. The synthesis of logical circuits can be affected one takes advantage of known techniques that allow the execution of one and two-qubit gates though the teleportation protocol [87].

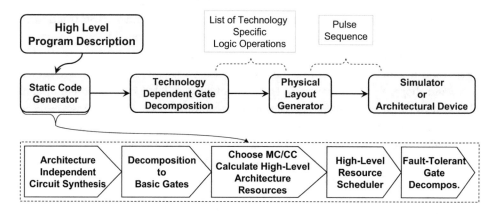

Figure 8.1: Compilation Layers

Figure 8.1 shows the components of an example quantum architecture compiler. A description of the program in a very general high-level manner such as a large unitary matrix or a high level C-like language such as QCL [152] that is technology- and layout-independent, serves as an input to the *Static Code Generator* (SCG). The bottom of Figure 8.1 shows the steps of the SCG component.

The first SCG stage performs technology and architecture independent circuit synthesis that breaks the algorithm into a useful, identifiable set of quantum operations. Once a universal set of basic gates is determined the SCG further decomposes the circuit into one and two-qubit gates, exposing all high-level qubit resources needed for the application. Next, the SCG determines appropriate error correcting codes for the memory and the computation, based on the architectural resource constraints and high-level structures. At this point, the program is composed of assembly language-like instructions as in Table 8.1, fully exposing the hardware resources and ready to be scheduled by the high-level scheduler. Finally, the SCG inserts fault-tolerance into each operation by decomposing each logical gate and LOAD/STORE operation into a fault-tolerant list of lower level quantum/classical instruction based on the choices for MC and CC.

The output of the Static Code Generator is a high-level quantum assembly language program that will describe a fault-tolerant, error correction enabled quantum algorithm with a clear description of the available quantum and classical resources at the system level. The next stage is the Technology Dependent Compiler (TDC), which knows nothing about the geometrical layout of individual tiles, but decomposes the quantum operations into the equivalent elementary operations available for the particular technology. In the case of ion-traps the output is a list of ion-trap logic gates consisting of single qubit rotations and controlled-Z gates.

Last is the Physical Layout Generator (LG), which takes in a description of the available resources and allocates physical locations and data-paths for each physical qubit in the system. The LG has full knowledge of the physical layout of each tile and schedules the elementary qubit operations accordingly, even if this changes the original sequence provided by the SCG. Assuming that maximum parallelism has been implemented at previous stages, the LG attempts to: 1) minimize the communication costs for multi-qubit gates at the physical level, and 2) optimize the physical resource distribution. The output of the LG is a sequence of control pulses fed into the device and ready for execution.

8.3 ARCHITECTURE-INDEPENDENT CIRCUIT SYNTHESIS

Architecture-independent circuit synthesis is analogous to the design and optimization of classical integrated circuits, where technology independent synthesis is performed using abstract logic gates. After this, the circuit is mapped to the technology by converting the gates to the gates best suited to the specified technology as described in Section 8.2. At any stage of the compilation flow (except perhaps the layout generator), given a general logical circuit, a compiler may identify various subcircuit structures which lend themselves to different optimization techniques.

At the highest possible level of a quantum algorithm, the action of the algorithm on n logical qubits is described as a $2^n \times 2^n$ unitary matrix. (This is analogous to using a boolean function describe a program in classical computation). Just as it is relatively easy to translate a boolean function into a corresponding circuit of operations, it is similarly possible to decompose an arbitrary n-qubit unitary operator into basic quantum logic gates. Also, exactly as in classical computation, the optimization of the resulting circuit is a nontrivial task. What may be truly different are the transformation rules for a quantum circuit.

There is a significant amount of ongoing work in architecture-independent quantum circuit synthesis. The first subcircuits that may be identified by a compiler are circuits composed entirely of controlled-NOT gates. Considerable work has been done in the area of synthesis of controlled^k-NOT circuits and general classical reversible circuits [104, 137, 159]. Transformation rules exist for controlled^k-NOT circuits (defined as NOT gates controlled by the AND of k bits). The transformation rules take any controlled^k-NOT circuit to an equivalent circuit in its canonical form, which can then be optimized using a heuristic whose cost function is the minimal average number of control qubits. Quantum modular exponentiation (the most expensive component of Shor's factoring algorithm) can be written entirely as a controlled^k-NOT circuit. Additionally, a technique for restructuring stabilizer circuits which are used in every error correction routine, is provided by Aaronson and Gottesman [1] Aaronson and Gottesman prove that any stabilizer circuit has an equivalent circuit in canonical form with only $O(n^2/\log n)$ gates, leaving open the question whether an optimal construction exists.

Generally, it is desirable to decompose any two-qubit operator into a number of controlled-NOT gates (i.e., CNOT) since the universal gate library [16] consists of CNOT gates and one-qubit gates. Song and Klappenecker [190] have devised a method for optimizing arbitrary controlled two-qubit operators. Shende, Bullock and Markov [180, 182, 183] propose tests and implement an algorithm that gives quantum circuits that simulate arbitrary two-qubit unitary operators. More specifically, they provide a method to determine which two-qubit operators are CNOT optimal, with the worst case being 3 CNOT gates. Other circuit synthesis work has been with arbitrary n-qubit diagonal operators [41, 97], and even incompletely specified two-qubit operators [180]. All the above mentioned circuit optimizations are implemented during the first stage of the quantum compiler example shown in Figure 8.1.

Although extensive groundwork has been done in architecture-independent circuit synthesis, a carefully designed quantum compiler can provide a framework from which to unify, refine, and expand these optimizations. In particular:

- A set of concrete transformation rules for stabilizer circuits can be given to provide a better canonical form than the one suggested in Reference [1]. The canonical form can be used to create heuristics that optimize error correction circuits.

- Creation of fault-tolerance preserving transformation rules that allows designers to change the universal set of basic gates while maintaining maximum time to failure in a high-level algorithm.

- The effects of combining different transformation rules has not been fully explored before. For example, it is unknown how the rules affect each other once they are implemented in common optimization tool.

8.4   MAPPING CIRCUITS TO ARCHITECTURE

While circuit synthesis is an important step, mapping these idealized circuits to a physical machine is perhaps the greatest opportunity for optimization. A custom hardware implementation of each circuit is not only impractical due to machine size constraints, but also technology-dependent elementary operations, large fault-tolerance overheads, and the use of teleportation all make the classical approach of direct circuit synthesis to hardware unappealing in the quantum domain. Creating schedulers that map quantum circuits onto equivalent fault-tolerant procedures that utilize the tradeoffs associated with the quantum hardware and possibilities at the systems level will be one of the key contributions of computer architects.

Let us consider an example, which describes part of the process of mapping and optimizing a controlled^k-NOT circuit. This is illustrated in Figure 8.2, which shows three equivalent controlled^k-

NOT based circuits. It is clearer to describe the circuit schematically rather than showing the instructions explicitly as in Table 8.1. When shown schematically one can "see" the needed communication from qubit to qubit when executing multi-qubit gates such as controlled operations.

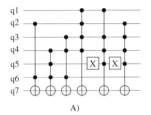
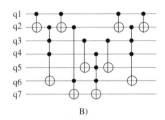
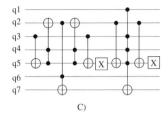

Figure 8.2: Optimizing circuits for fewer control bits. The circles are NOT gates controlled by the AND of the qubits connected by solid dots on the vertical lines. The boxes marked with an "X" are bit-flip gates. We want to minimize the number of solid dots per gate without dramatically increasing the number of gates. (A) An unoptimized Controlled-Not based circuit in its canonical form [104]. (B) Using the transformation rules in [104] to reduce the number of control qubits per gate to 1.7. (C) Using boolean algebra simplification of the circuit in (A) to reach 1.4 control qubits per gate.

The circuit in Figure 8.2(a) is the derived canonical form using the transformation rules provided in [104] of an initial unoptimized CNOTbased circuit. The canonical form is a useful starting point for the optimization of any boolean CNOTbased circuit since it allows all NOT operations to be concentrated on the last qubit ($q7$). This means that the gates can be executed in any order and it would be straight forward to apply boolean algebra simplification. A good CNOTbased circuit cost metric is the minimization of the number of control qubits per gate, which is what the authors of [104] strive for. Their result is shown in Figure 8.2(b). This is a reasonable assumption since any controlled^k-NOT gate with $k > 2$ must be divided into $(2 * \lfloor \log k \rfloor + 1)$ *Toffoli* gates (a 3-input, 3-output reversible NAND gate, implemented as a NOT with two control bits), using $\lfloor \log k \rfloor$ additional logical ancilla qubits adding to the overall resources needed from the architecture. In addition, the synthesis of each Toffoli gate into one and two qubit operations is relatively expensive: a Toffoli gate divides into 2 Hadamard gates, 1 S gate, 6 CNOT gates, and 7 T gates [149] as in Figure 2.4. The T gates are essentially two-qubit gates since they require interaction with the specialized ancilla qubits and need both accumulators in a PE. An even better circuit in terms of the control-qubit cost is Figure 8.2(c), which we derived using simplification rules derived from boolean algebra (i.e., $A \oplus BA = A\overline{B}$).

Ideally, quantum researchers would like to create a compiler that can recognize the optimal circuit structure of Figure 8.2 for the specified architectural constraints. The compiler will choose the MC and CC encodings, decompose the basic circuit gates into fault-tolerant procedures, each individually optimized over the architecture such that the cost of communication, computation, and classical resource overhead throughout the high-level circuit execution is minimized. In a circuit of one and two-qubit gates, the most expensive operation is a transfer of a logical qubit from the MC

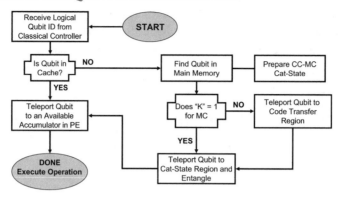

Figure 8.3: Overview of cache read operation used to estimate memory read time.

encoded memory to the CC encoded cache or PE. The *memory read* time (MR_t) roughly estimated from Figure 8.3 is equal to two MC teleportation steps, five MC error corrections, five logical gate times over MC whose sequence diagram is shown in Figure 6.4, and one error correction step over CC. A cache-miss is equal to MR_t; however, a cache-hit is just a single teleportation step from the cache to the processing element over the CC encoding. Thus, the cache read time (CR_t) is therefore,

$$CR_t = X(t_{tel,CC}) + Y(MR_t),$$

which is a weighted average of the expected cache-hit rate versus the expected cache-miss rate. The CC is chosen such that the time for a logical operation after the retrieval of the data should be much less than the time for an operation over an MC encoded qubit, where the tradeoff is that the logical qubits stored over MC are much more reliable.

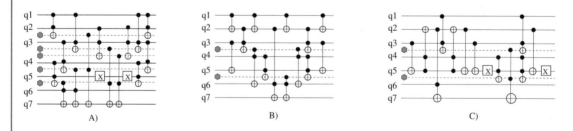

Figure 8.4: Equivalent circuits to the ones in Figure 8.2, but with all gates decomposed into Toffoli gates and full exposure of the logical qubit resources required. The dotted horizontal lines represent ancilla qubits, which are added to reduce overall communication costs once mapped to the architecture. Network (a) is least desirable, using the most logical qubits. Network (c) is the most desirable as it decomposes into fewer elementary operations.

Figure 8.4 shows the circuits from Figure 8.2 with all controlled$^{k>2}$-NOT gates broken into CNOT and Toffoli's only using a number of auxiliary logical qubits initialized at the "0" encoded state (dashed lines). The gates are reordered to expose the available parallelism at this level; however, each Toffoli gate is not yet decomposed into one- and two-qubit gates as shown in Figure 2.4. Each gate is a fault-tolerant logical gate composed of a number of physical operations over the PE tile whose physical circuit depth and dimensions are determined by the choice of CC. One can imagine the magnitude of the computation even for such a small circuit. Without explicitly calculating the schedule for each circuit over basic single and two-qubit operations, we see that the circuit in Figure 8.4(a) uses three more logical qubits than both other circuits. In addition, it requires 10 Toffoli gate time-steps over 15 total Toffoli gates. Figures 8.4(b) and 8.4(c) require only 7 and 6 Toffoli time-steps and 8 and 7 total Toffoli gates, respectively. The number of Toffoli timesteps limits the circuit's overall performance making the circuit in Figure 8.4(a) least desirable, even with infinite resources and parallelism. The above, however, does not provide us with much information about the circuit's communication costs without more careful consideration. To avoid any cache misses, a Toffoli gate will require two PE units and two logical qubit cache tiles. The time for a Toffoli gate will be roughly equal to the time of 14 logical operations over CC, plus the starting cost of loading the three data qubits from memory.

The controlled^k-NOT circuit example provides only an overview of the complexity involved in implementing just one portion of the compiler design flow shown in Figure 8.1 - the static code generator. The most important first stages of a quantum compiler implementation must be to identify the unique elements of quantum computing circuits and properties of quantum computation that will allow us to create the corresponding intermediate compiler data structures. Aside from tracing the necessary intermediate steps of a quantum compiler, we can identify several important challenges for system designers when designing a compiler for the development of large-scale quantum applications. These include the following:

- The development of simulation and modeling techniques for the quantum circuits involved in the implementation of the high-level architectural elements.

- Finding suitable cost metrics for compiler optimization that will allow us to generate and evaluate efficient fault-tolerant circuits at both the architecture level and the physical level of execution for a given quantum application.

- Identifying algorithms that insert, preserve, and optimize low-level, fault-tolerant circuits that implement high-level computations.

- Devising architectural strategies that can exploit uniquely quantum computation and communication resources, such as teleportation-based error correction and varying logically universal sets of quantum operations. In Chapter 10 we describe the possibility of teleportation-based quantum operations and error-correction.

8.5   OPTIMIZATION OF THE LOGICAL QUBIT TILES

The QLA architecture and other system-level studies [15, 78, 121, 154, 213] offer specific design points that exhibit efficient global communication and high-level organization. While these are critical to the scalability of a quantum computer, the careful design of the basic building blocks of the microarchitecture is also critical. In this chapter, we consider the effects of a quantum computer compiler on the performance parameters of the logical qubit tiles. We discuss the tradeoffs associated with the interaction of the quantum and classical parts in a quantum computer and consider answers to two critical questions: (1) is the low-level device architecture feasible when we take into account a more realistic technology model, and (2) how does the required control circuitry affect the execution and fault-tolerance of quantum circuits?

8.5.1   THE FAULT-TOLERANT THRESHOLD ESTIMATES

The above questions are considered regarding the communication requirements of the logical CNOT gate quantum circuit mapped on to an ion-trap device using the scheduler QPOS [141]. We use the threshold calculation methodology and circuit constructions outlined in Reference [201], where the authors consider a two-dimensional nearest-neighbor lattice architecture suitable for technologies with fixed qubits and nearest-neighbor only interactions [93, 98, 109]. The ion-trap ballistic communication model can be made analogous to nearest-neighbor communication by replacing each SWAP gate with a MOVE operation from one zone to the next.

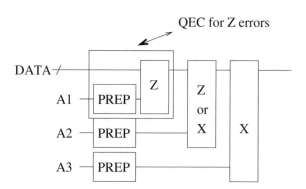

Figure 8.5: A schematic for a single deterministic QEC procedure for the Steane [[7, 1, 3]] code.

The authors in Reference [201] calculate a lower-bound for the physical threshold failure rate for the logical CNOT gate using the Steane [[7, 1, 3]] error correcting code to be on the order of $O(10^{-5})$ by assuming an adversarial stochastic noise model, where every physical gate location in the logical CNOT circuit (i.e., each of the CNOT gates, wait gates, movement gates, and measurement gates) can fail with some probability associated with the gate type. We find that, if all gates are treated equally, the threshold for the logical CNOT circuit is approximately 3.1×10^{-6}, and further investigate the threshold as we vary the failure rate of each type of location relative to the failure rates of the rest of the location types.

The logical CNOT gate is the most computationally intensive gate in logical quantum circuits, as it requires moving either the logical control qubit or logical target qubit so they can interact, and is then followed by error correction on both qubits. Thus, the component threshold failure rate for other logical gates will be necessarily higher than the one for the logical CNOT gate. To estimate the threshold value of the CNOT circuit, we employ the methodology described in rigorous detail in References [7] and [201], where the number M of *malignant* pairs of locations in the circuit are counted. A malignant pair of gates within a fault-tolerant circuit is any gate pair that causes more than one error at the output. If the circuit is fault-tolerant, then no single gate will cause the circuit to fail; however, many pairs of gates may cause it to fail. In some cases the errors cancel one another. The threshold value p_{th} is then lower-bounded by $p_{th} \geq M^{-1}$, where the authors in Reference [201] demonstrate that this is a tight lower bound, and thus a relatively reliable estimate.

The pair counting method requires that we use a deterministic syndrome extraction approach for the $[[7, 1, 3]]$ QEC circuit shown in Figure 8.5. Three ancilla qubits $\{A1, A2, A3\}$ are prepared using the circuit shown in Figure 4.8 in Chapter 4, such that they are available for the interaction with the data. If at least two ancilla pass the preparation procedure, then the X and Z syndromes can be extracted. If the first ancilla passes, then the Z syndrome is extracted with it. If the first ancilla doesn't pass, but the second one does, then the Z syndrome is extracted with the second ancilla. Similarly, the X syndrome uses the second and third ancilla blocks. If two or more ancilla blocks fail preparation, it is assumed that logical error is transferred to the data because there would be no good ancilla left to correct it. What makes this QEC circuit fault-tolerant is that a single error in the ancilla preparation may cause the preparation to fail still leaving us with two good ancillas. For an error to be passed to the data from a faulty preparation stage, we need two errors to happen in two different ancilla blocks, which is a second order event.

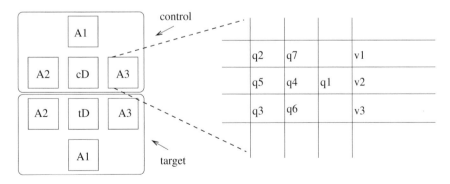

Figure 8.6: The physical layout description for the logical CNOTgate.

Figure 8.6 shows the layout for the entire logical CNOT gate. The control and the target data blocks are labeled as cD and tD, each of which is composed of seven lower-level qubits encoded with the Steane [[7, 1, 3]] error correcting code. Each data block is given three ancilla blocks for a deterministic implementation of the circuit. On the right-hand-side of the figure is shown the physical qubit arrangement for each logical sub-block in the CNOT gate. Ions $\{q1 \ to \ q7\}$ are arranged to allow the least movement possible for the gates needed to prepare the ancilla. The rule of thumb we employ to generate an efficient 2-D layout for the encoder is as follows: (1) we generate by hand a communication density graph (CDG) for each circuit, where each ion marks a vertex and each physical CNOT gate denotes an edge between two vertices, and (2) we then create a 2-D CDG visual description with the least amount of intersecting edges.

To ensure that the physical layout for each qubit block is truly efficient, we generated 1000 random mappings with the scheduler. For each one, we place the qubits randomly on the 2-D ion-trap grid for the encoder. We found, that each of the 1000 random schedules had more cycles than the schedule produced when starting with the specified ion locations in Figure 8.6. We placed the ions on the 2-D grid shown on the right-hand-side of Figure 8.6, such that each qubit moves on only one or two Junctions for the entire ancilla preparation procedure.

8.5.2   CIRCUIT SCHEDULING AND THE FAULT-TOLERANCE CONSTRAINT

To schedule the CNOT circuit, we use our scheduling heuristic, QPOS [141], described in Section 8.1. The original QPOS implementation is modified to allow a more accurate scheduling of the pulse sequences necessary for the ion-trap technology, using the trap abstraction shown in Figure 8.7. Figure 8.7a depicts a single trap zone consisting of three electrodes laid out in a planer fashion on either side of the trapping region, as demonstrated in recent ion-trap literature [178].

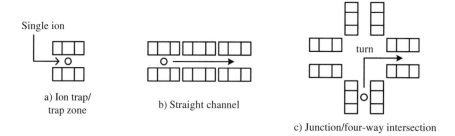

Figure 8.7: a) Our abstract trap zone, which is capable of generating the potential well to trap an ion, consists of three electrodes on either side of a vacuum. b) Trap zones may be laid in sequence to form a straight channel for ballistic movement. c) A junction consists of four neighboring trap zones. Electrodes from all four are required to perform a move through the junction.

A straight channel for qubit movement is created by lining up a sequence of such trap zones, as shown in Figure 8.7b. A positively charged ion is then ballistically transported by lowering the potential of electrodes "in front" of it and raising the potential of electrodes "behind" it. The details of the exact pulse sequences are necessarily more complicated than this, in order to more carefully manage the evolution of the potential well. However, by defining our trap zones to consist of three electrodes on either side, we may abstract away the complexity of the full sequences and simply assume that movement between adjacent trapping zones utilizes only the electrodes in these two zones. Finally, we need junctions to allow routing of qubits in two dimensions. Figure 8.7c shows an abstraction of a four-way intersection, where four trap zones may communicate. The movement of a qubit through such a junction requires the application of specific pulse sequences to electrodes in all four adjacent trap zones.

To evaluate the resource requirements for fault tolerant logical blocks such as the logical CNOT gate, each operation in the assembly-level circuit description is decomposed into the required cycles of operation that correspond to the duration of the gate in the ion-trap technology. In the scheduled circuit program, we have fixed all gate durations to be relative to the duration of a single-qubit operation (i.e., 1 μs), which is defined as 1 clock cycle. This yields the following physical circuit locations and their decomposition in the computed schedule:

- MOVE ION Z1 -> Z2: where Z1 and Z2 denote two adjacent trap zones. At approximately 30 m/s ion speed [169], and $\approx$ 100 μm per trap zone of three control electrodes [178], the MOVE operation in a straight line requires 7 cycles to complete, and across a junction 14 cycles.

- MEASURE ION: A measurement operation which we assume will require 10 cycles. The current measurement time is approximately 100 μs — however, 10 μs is a possibility [130].

- JOIN/SPLIT ION1,ION2: As suggested in the literature [193], joining and splitting two ions in adjacent trap zones is not expected to take more than a few microseconds. We assume a duration of 1 cycle each.

- CNOT ION1,ION2: In recent ion-trap experiments [18, 169, 171], the CNOT gate has been implemented using the following sequence: 1-qubit gate, join, Controlled-Phase Gate (8 cycles), split, 1-qubit gate, join, Controlled-Phase Gate (8 cycles), split, 1-qubit gate, for a total assumed duration of 23 cycles.

- WAIT ION: An ion can acquire memory errors while standing idle (not having any operation performed on it during a cycle). The experimentally verified decoherence time of ions is 14.3 seconds [130]; however, the decoherence time is expected to reach $50 - 100$ seconds.

The 7-bit encoder circuit is particularly interesting, because the nine CNOT gates in the encoder can be executed in any order without affecting the circuit functionality. However, the order of the gates causes errors to propagate differently and thus, the *verification* circuit portion may fail. This is what we call *the fault-tolerance constraint*. To demonstrate the fault-tolerance constraint when scheduling quantum programs, we have labeled the 9 encoding CNOT gates "*a*" through "*i*" in Figure 8.8.

The original gate order in the encoder circuit (as shown in Figure 4.8) allows the propagation of bit-flip errors such that any non-trivial error pattern (i.e., two or more errors) can be detected by observing the parity of the bits obtained when measuring only qubits $q2$, $q4$, and $q6$. The circuit in

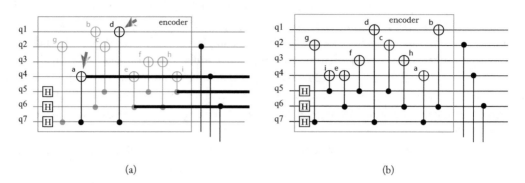

(a)   (b)

Figure 8.8: (a) The error propagation pattern if the control bit of gate "a" fails. We see that the output error enters two of the three verification gates, causing the verifier to yield even parity in the measurement of the verification qubits, and thus a "no-error" result. (b) The encoder circuit with the gates completely reordered, but still fault-tolerant (coincidentally, gate "g" is still scheduled first). In an error-free environment, the functionality of the encoding circuit is preserved through any permutation of the gates shown.

Figure 8.8(a) shows the encoding circuit where gate "g" has been scheduled first. In this manner, a *single failure* at either gates "a" or "d" will cause more than one error to pass through the verification circuit and cause the verifier to return an even parity result and thus ("no-error"), even though there are multiple errors on the encoded data. In the example shown, the error will propagate to both qubits $q4$ and $q6$ at the output of the circuit. Thus, a schedule where a single failure in the circuit causes greater than one error at the output means that the fault-tolerance constraint required by the Steane $[[7, 1, 3]]$ error correcting code is violated. The Steane error correcting code can only correct a single error. Therefore, a fault-tolerant encoder must ensure that at any one error in the circuit causes at most one error in the output, and if it causes more, it must be detected by the verification circuit.

At first glance, the circuit in Figure 8.8(b) looks the same as the one in Figure 8.8(a) with gate "g" still scheduled first (this was accidental by our scheduler); however, in this ordering, there is nowhere in the circuit where a single error will cause the verifier to yield an incorrect result. In fact, the circuit in Figure 8.8(b) can be used instead of the one in Figure 4.8.

When calculating gate dependencies based on the fault-tolerance constraint, we found it prohibitive to evaluate the propagation of errors at each step of the QPOS scheduler; Therefore, to satisfy the dependency constraint, we dynamically linked the scheduler to the QASM-TOOLS suite [59] that indicates the existence of any single error in a given circuit that causes greater than one error at the output. For general quantum circuits, we allowed the input circuit program to indicate via a flag if the fault-tolerance constraint is necessary. The traditional encoder and verifier for the Steane $[[7, 1, 3]]$ code [7], for example, is not bound by the fault-tolerance constraint; however, it also requires more gates and 14 qubits instead of 10.

The fault-tolerance constraint is an important dependency condition for quantum operations that has not been considered anywhere else in the quantum compilation literature. It highlights the difference in compiling quantum programs vs. classical programs: two operations may not have data dependencies, but dependencies can be introduced through fault tolerant error correction requirements. Having a scheduler that schedules quantum operations and ensures that the constraint is satisfied is necessary in order to accurately study fault tolerant LCBs and the interactions of each of the LCB components. Entanglement aids the spread of errors so quickly that, even if the smallest most innocuous low-level LCB is not fault-tolerant, then no amount of higher-level error correction will improve the reliability of quantum operations (and no useful quantum computing can happen).

Another roadblock faced when trying to implement the scheduling mechanism for the ion-trap technology is the difficulty caused by the quantum computing paradigm when creating an optimal geometrical layout for maximum SIMD gate execution. Maximizing SIMD execution will alleviate some of the classical resource overhead (identified as a bottleneck [114]) in ion-trap chips by making each electric pulse apply to as many gates as possible during any given program cycle. The problem, however, is that different ions cannot be reassigned to different operations when such an opportunity is identified. For example, the three verification qubits $\{v1, v2, v3\}$ can be assigned to any of the three data qubits $\{q2, q4, q6\}$; however, the scheduler cannot dynamically recognize this and thus prioritize the communication resources accordingly because it cannot track how each qubit is entangled with other qubits through previous stages of the program. In other words, *register renaming* is not possible for quantum programs, where $A = B$, $B = C$, but because of entanglement $A \neq C$.

8.5.3   THRESHOLD CALCULATIONS

We estimated the threshold value for a logical CNOT block with the assumption that each location of any gate type is treated simply as single location that takes exactly one cycle of execution and there is some probability that it will introduce an error on the qubits it acts on. The results are shown in Figure 8.9. Figure 8.9(a) shows the circuit statistics, which can be divided intro three distinct subroutines: encoding, logical CNOT gate, and error correction procedure. We see that, after mapping the communication paths, the circuit is inherently serial and there is much more idling (e.g., WAIT gates) than any other operation. The total time cycles of the circuit for the logical CNOT gate is 73, so if a single time cycle is upper bounded by the duration of a two-qubit gate (i.e., $7 \mu s$) the total duration of a logical CNOT gate at a single level of recursion can be as much as 500 microseconds (for an ion-trap device). This is well within the decoherence time of 14.3 seconds for individual ions.

Since each location (or each logic gate) in a quantum device requires separate physical calibration (and even mechanisms), we consider how the threshold is affected as the relative reliability of each gate is varied relative to other gate-types. The results are shown in Figure 8.9(b), where it can be seen that the gates in the circuit that affect the threshold value the most are "WAIT" gates and "MOVE" gates. In fact, when "WAIT" gates are set to be 100 times more reliable than the rest

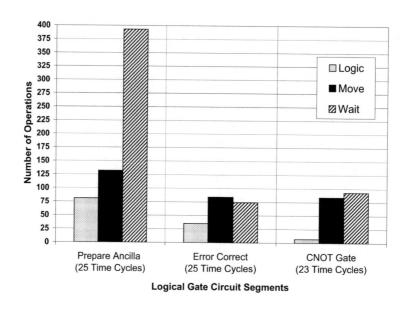

(a)

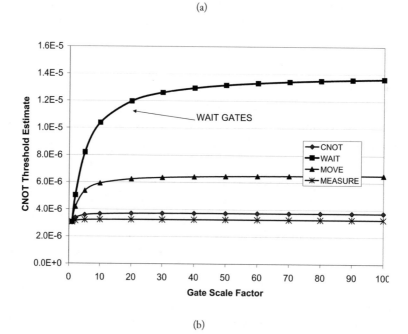

(b)

Figure 8.9: Caption on the next page.

Figure 8.9: (a) The total operation count for the logical CNOT circuit when using the QPOS heuristic scheduler. Each operation is assumed to take exactly one cycle of duration, and all the logic gates such as CNOT gates, single-qubit gates, and measurement gates have been grouped under the term "logic". **(b)** Variance of the threshold for the logical CNOT circuit as a function of the probability of gate failure relative to the rest of the gate-types. We see that the threshold improves by an order magnitude (to 1.36×10^{-5}) if WAIT gates are 100 times more reliable than the rest of the gates in the circuit. The threshold is much less affected by changing the relative weight of the other gate-types to the rest of the locations.

of the gates in the circuit, the threshold improves by almost an order of magnitude (to approximately 1.4×10^{-5}). There are a total of 58700 malignant pairs where one of them is a CNOT gate, 19792 pairs of locations where one them includes a measurement operation, 174076 pairs with a "MOVE" operation, and 254909 pairs with a "WAIT" operation (see Figure 8.9(b)).

8.5.4 SUMMARY DISCUSSION

As can be seen from the above analysis, the communication latencies inherent in a 2-D physical device for the ion-trap technology significantly limit the scalability of the devices by lowering the threshold value (lower threshold value means that recursion level must increase to achieve desired system reliability). Optimistic ion-trap parameters are still below the lower-bound threshold estimate of 3.1×10^{-6} — however, they may have to be even lower for long computations that require the manipulations of millions of logical qubits through millions of logical gates.

Furthermore, the model shown here, where single time cycle for each operation is assumed, is at best optimistic for the ion-trap technology. A more realistic model is to consider the pulse sequences necessary for each gate type. For example, moving between adjacent trap zones requires moving through a distance of almost 200 micrometers, which means that the straight movement operation takes approximately 7 μs. A measurement operation requires 10 μs and a CNOT gate requires two controlled-Phase gates combined with 3 single-qubit gates, and joining and splitting of the ions, for a total of approximately 23 μs per CNOT gate. Finally, moving through a junction is more expensive than moving between two adjacent trap zones (based on our discussions with NIST researchers it is estimated to be twice as slow). Therefore, if a single cycle is considered to be the duration of the fastest gate, such as single-qubit gates, and the schedule of the circuit is remapped to reflect the varying duration for each gate-type, the lower-bound of the threshold value for ion-traps decreases substantially.

We performed preliminary estimates of how the threshold is affected by changing the scheduling heuristic to reflect the varying gate latencies. We defined a basic program cycle to be 1 μs, which is the duration of a single-qubit gate. Thus, a single movement operation requires that the ions moving undergo 7 WAIT gates, and a single CNOT operation requires two two-qubit gates, three single-qubit gates on the affected ion. The new threshold estimates for this more realistic ion-trap model are shown in Figure 8.10. This is not the threshold; however, for the entire logical CNOT circuit, this is the threshold for only a single QEC circuit (a very small portion of the CNOT circuit). Thus, we expect that the threshold value for the CNOT circuit to be even lower.

From our rough estimate of the single QEC circuit with the new device assumptions, we see that memory errors have an enormous impact on the device scalability. The QLA architecture was modeled by assuming ion lifetime of 50 seconds or more, which nearly negates the effect of memory errors; however, in the analysis provided here, experimental lifetimes of 14.3 seconds are considered. The threshold when all gate cycle are treated equally is estimated to be 1.8×10^{-8} and improves two orders of magnitude to 1.7×10^{-6} when WAIT gates are 100 times more reliable than other gates. The threshold seems to be affected very little by any of the other gate types when wait gates are counted.

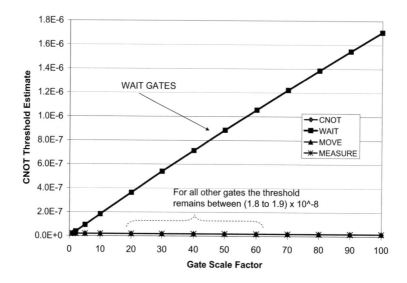

Figure 8.10: Variance of the threshold for the logical CNOT circuit as a function of the probability of gate failure relative to the rest of the gate-types with the more realistic ion-trap model. The threshold when all gate cycle are treated equally is estimated to be 1.8×10^{-8} and improves to 1.7×10^{-6} when WAIT gates are 100 times better; however, it is not affected by making any of the other gates better.

Our estimate of the threshold value of ion-trap devices is reflected by our abstraction of the devices themselves. Our abstractions, however, are general enough to suggest that significant improvements in the technology are required for true scalability to be achieved. On the other hand, we find that the abstractions we make for any high-level ion-trap device play a significant role in our ability to estimate the scalability of the technology. To design a large-scale quantum computing system that accurately reflects the capabilities of an assumed technology, it is important to create a simulation and device optimization infrastructure that not only takes into account a theoretical estimate of the threshold value, but also the effect of the interaction of the classical and quantum components in the devices on the failure rate of the logical operations.

In addition to improvements in the technology, we have not considered improvements in the error correction circuits themselves, which will further increase the threshold value estimate. For example, the schedule produced by QPOS for the preparations circuits was limited by the fact that the schedule only included *legal* gate reordering that preserved the fault-tolerant properties of the preparation circuit. It would be more interesting if we can find the optimal gate order that gives us the best possible communication paths and automatically generates the verification circuit needed to make the entire circuit fault-tolerant. Furthermore, the natural two-qubit gate for ion-traps is the controlled-Phase gate, and we expect that the implementation of the $[[7, 1, 3]]$ circuit will improve considerably if we replace all gates with controlled-Phase gates.

CHAPTER 9

Using the QLA for Quantum Simulation: The Transverse Ising Model

In this chapter, we describe the usage of the QLA to implement an important quantum application that is not factoring: quantum simulation. More specifically, we summarize the work of Clark et al. [54], which estimates the resource requirements for a quantum simulation of the ground state energy for the 1-D quantum Transverse Ising Model, incorporating the fault-tolerant quantum error correction techniques employed by the QLA. The authors (which include one of the present book authors) apply the general approach of Abrams and Lloyd [2, 3], and compute estimates for the total number of physical qubits and computational time as a function of the number of particles (N) and required numerical precision (M) in the estimate of the ground state energy. The reason for choosing the 1-D quantum TIM model is because it is well studied in the literature and has an analytical solution that can be computed classically [107, 160, 175]. This makes future validation studies possible.

9.1 THE TRANSVERSE ISING MODEL OVERVIEW

The 1-D Transverse Ising Model is one of the simplest models exhibiting a quantum phase transition at zero temperature [73, 106, 160, 175]. The calculation of the ground state energy of the TIM varies from analytically solvable in the linear case [160] to computationally inefficient for frustrated 2-D lattices [144]. For example, the calculation of the magnetic behavior of frustrated Ising antiferromagnets requires computationally intensive Monte-Carlo simulations [100]. Given the difficulty of the generic problem and the centrality of the TIM to studies of quantum phase transitions and quantum annealing, the TIM is a good benchmark model for quantum computation studies.

The Transverse Ising Model consists of N-spin-1/2 particles with nearest-neighbor spin-spin interactions along the z-axis in the presence of an external magnetic field along the x-axis. The Hamiltonian, H_I, for this system is:

$$H_I = \sum_i \Gamma \sigma_i^x + \sum_{\langle i,j \rangle} J_{ij} \sigma_i^z \sigma_j^z, \qquad (9.1)$$

where J is the spin-spin interaction energy, Γ is the energy of a spin the external magnetic field, and $\langle i, j \rangle$ implies a sum only over nearest-neighbors [175]. σ_i^x and σ_i^z are the Pauli spin operators for the ith spin, and we set $\hbar = 1$ throughout this paper.

This chapter focuses on the 1-D linear chain TIM of N-spins with constant Ising interaction energy $J_{ij} = -J$. The ground state of the system is determined by the ratio of $g = \Gamma/J$. For the large magnetic field case, $g >> 1$ the system is paramagnetic with all the spins aligned along the $\hat{x}$ axis, and in the limit of small magnetic field, $g << 1$, the system has two degenerate ferromagnetic ground states, parallel and anti-parallel to the $\hat{z}$ axis. In the intermediate range of magnetic field strength the linear 1-D TIM exhibits a quantum phase transition at $g = 1$ [175].

The TIM Hamiltonian in Equation (9.1), for the 1-D case with constant coupling can be rewritten as:

$$H_I = -J \left(\sum_{j=1}^{N} g X_j + \sum_{j=1}^{N-1} Z_j Z_{j+1} \right) \tag{9.2}$$

where the Pauli spin operators, σ_j^x and σ_j^z, are replaced with their corresponding matrix operators X_j and Z_j. For the 1-D TIM, the ground state energy can be calculated analytically in the limit of large N [160]. In the case of a finite number of spins with non-uniform spin-spin interactions (J not constant), it is possible to efficiently simulate the TIM using either the Monte-Carlo method [176] or the density matrix renormalization group approach [107]. The challenge for classical computers comes from the 2-D TIM on a frustrated lattice where the simulation scales exponentially with N. Applying the quantum phase estimation circuit to calculate the ground state energy of the TIM requires physical qubit resources, which scale polynomially with N, and the number of computational time steps is also polynomial in N, although exponential in the numerical precision M. In addition, just as the complexity of the problem is independent of the lattice dimension and layout when applying classical brute force diagonalization, the amount of resources required to apply the quantum phase estimation circuit is largely independent of the dimensionality of the TIM Hamiltonian.

9.2   TIM QUANTUM SIMULATION RESOURCE ESTIMATES

The approach of Clark et al. [54] to estimating the resource requirements for the TIM ground-state energy calculation with Hamiltonian H_I involves two steps. First, the authors follow the approach of Abrams and Lloyd and map the problem of computing the eigenvalues of the TIM Hamiltonian in Equation (9.2) onto a phase estimation quantum circuit [2, 3]. Second, the authors decompose each operation in the phase estimation circuit into a set of universal gates that can be implemented fault-tolerantly within the context of the QLA architecture. This allows for an accurate estimate of the resources in a fault-tolerant environment.

9.2.1   PHASE ESTIMATION CIRCUIT

The phase estimation algorithm calculates an M-bit estimate of the phase ϕ of the eigenvalue $e^{-i2\pi\phi}$ of the time evolution unitary operator $U(\tau) = e^{-iH_I\tau}$ for fixed τ, given an eigenvector of H_I. $\phi < 1$ and can be represented by the binary fraction $0.x_1 \ldots x_M$ [2, 3]. The energy eigenvalue $E = \frac{2\pi\phi}{\tau}$ when $E\tau < 2\pi$. Calculation of the ground state energy $|E_g|$ requires that $\tau < 2\pi/|E_g|$. For the 1-D TIM, the magnitude of the ground-state energy $|E_g|$ is bounded by $NJ(1+g)$ [160]. In the region near the phase transition $g \approx 1$, we choose $\tau = (10JN)^{-1}$.

The quantum circuit for implementing the phase estimation algorithm is shown in Figure 9.1. The circuit consists of two quantum registers: an N-qubit input quantum register prepared in an initial quantum state $|\Psi\rangle$, and an output quantum register consisting of a single qubit recycled M times [55, 158]. Each of the N qubits in the input register corresponds to one of the N spin-1/2 particles in the TIM model [39]. At the beginning of each of the M steps in the algorithm, the output qubit is prepared into the state $\frac{1}{\sqrt{2}}(|0\rangle + |1\rangle)$ using a Hadamard (H) gate. The H gate is followed by a controlled power of $U(\tau)$, denoted with $U(2^m\tau)$, applied on the input register, where $0 \leq m \leq M - 1$.

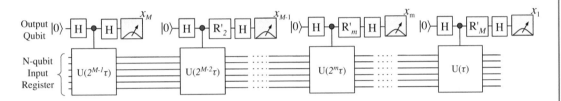

Figure 9.1: The circuit for implementing the phase estimation algorithm using one continuously recycled control qubit.

Letting j denote to the jth step in the circuit, each time the output qubit is measured (meter symbols), the result is in the mth bit in the estimate of ϕ, following the rotation of the output qubit via the gate:

$$R_j = |0\rangle\langle 0| + \exp\left(i\pi \sum_{m=M+2-j}^{M} \frac{2^{M+1}x_m}{2^{m+j}}\right)|1\rangle\langle 1| \qquad (9.3)$$

where the gate R_j corresponds to the application of the Quantum Fourier Transform on the output qubit at each step [55, 158]. The result after each of the M measurements is an M-bit binary string $\{x_1x_2\ldots x_M\}$, which corresponds the M-bit approximation of ϕ given by $0.x_1 \ldots x_M$. Using this estimate of ϕ, the corresponding energy eigenvalue $E = \frac{2\pi\phi}{\tau}$ will be the ground-state energy E_g with probability equal to $|\langle\Psi|\Psi_g\rangle|^2$ [2], where $|\Psi_g\rangle$ is the ground eigenstate of H_I.

To maximize the probability of success $|\langle\Psi|\Psi_g\rangle|^2$, the initial quantum state $|\Psi\rangle$ should be an approximation of the ground state $|\Psi_g\rangle$. For arbitrary Hamiltonians, the preparation of an approximation to $|\Psi_g\rangle$ is generally computationally difficult [27, 111]. For certain cases, the preparation can be accomplished using classical approximation techniques to calculate an estimated wavefunction or adiabatic quantum state preparation techniques [10, 39]. If the state can be prepared adiabatically, the resource requirements for preparing $|\Psi\rangle$ are comparable in complexity to the resource requirements for implementing the circuit for the phase estimation algorithm shown in Figure 9.1 [39]. For this reason, focus is given on estimating the number of computational time steps and qubits required to implement the circuit, assuming that the input register has been already prepared in the N-qubit quantum state $|\Psi\rangle$.

9.2.2 DECOMPOSITION OF THE TIM QUANTUM CIRCUIT INTO FAULT-TOLERANT GATES

Figure 9.1 in Section 9.2.1 shows the TIM circuit at a high-level, involving $N + 1$ unitary operators. In this section, each unitary operation of the circuit is decomposed into a set of basic one and two qubit gates, which can be implemented fault-tolerantly using the QLA architecture. The high-level circuit operations which require decomposition are the controlled-$U(2^m\tau)$ gates and each R_j gate.

The Controlled-$U(2^m\tau)$ gate can be decomposed using the second-order Trotter formula [149, 198]. First, H_I is broken into two terms: $H_X = \sum_{j=0}^{N} gX_j$, representing the transverse magnetic field, and $H_{ZZ} = \sum_{j=0}^{N-1} Z_j Z_{j+1}$, representing the Ising interactions. By considering the related unitary operators

$$U_x(2\tau) = \prod_{j=1}^{N} \exp(-ig\tau X_j) \qquad (9.4)$$

$$U_{zz}(2\tau) = \prod_{j=1}^{N-1} \exp(-i\tau Z_j Z_{j+1}), \qquad (9.5)$$

and setting $g = 1$ (as discussed in Section 9.1), we can construct the Totter approximation of $U(2^m\tau)$, denoted by $\tilde{U}(2^m\tau)$ as:

$$\begin{aligned} U(2^m\tau) &= \left[U_x(\theta)\, U_{zz}(2\theta)\, U_x(\theta)\right]^k + \epsilon_T \\ &= \tilde{U}(2^m\tau) + \epsilon_T, \end{aligned} \qquad (9.6)$$

where $\theta = (2^m\tau/k)$ and ϵ_T is the Trotter approximation error, which scales as $\mathcal{O}\left(\frac{(2^m\tau)^3}{k^2}\right)$ [198]. The Trotter approximation error can be made arbitrarily small by increasing the integer Trotter parameter k. Since the controlled-$U(2^m\tau)$ corresponds to the $(M - m)$th bit, ϵ_T must be less than $1/2^{M-m}$, which is the precision of the $(M - m)$th measured bit in the binary fraction for the phase ϕ. Thus, when approximating $U(2^m\tau)$, k is increased until ϵ_T is less than $1/2^{M-m}$. For a given M, the authors estimate a numerical value for the Trotter parameter $k(m = 0) = k_0$ as a function of $N \leq 10$, with the constraint that $\epsilon_T < 1/2^M$. Clark et al. [54] thus find that for fixed M, k_0 scales as $1/N$, where k_0 is extrapolated for larger N based on a power-law fit of $N \leq 10$. For $m > 0$, Clark et al. [54] set $k = 2^m k_0$, which will satisfy the error bound based on the scaling of ϵ_T with k.

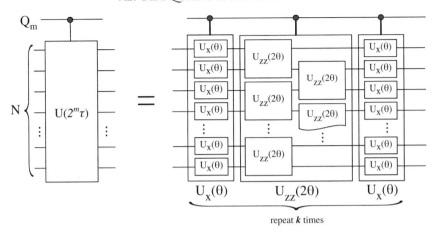

Figure 9.2: Circuit for the controlled unitary operation $U(2^m \tau)$ approximated using the Trotter formula.

The circuit corresponding to the Trotter approximation of $U(2^m \tau)$ is shown in Figure 9.2, where it can be seen that the controlled-$U(2^m \tau)$ is composed of two controlled-$U_x(\theta)$ operations and a controlled-$U_{zz}(\theta)$ operation, repeated k times and controlled on the mth instance of the output qubit denoted with Q_m. Expanding the circuit in Figure 9.2, Clark et al. [54] show that $\tilde{U}(2^m \tau)$ can be expressed as:

$$\tilde{U}(2^m \tau) = U_x(\theta) \left[U_{zz}(2\theta) U_x(2\theta) \right]^{k-1} U_{zz}(2\theta) U_x(\theta), \tag{9.7}$$

which shows that, approximating $U(2^m \tau)$ will require the sequential implementation of k controlled-$U_{zz}(2\theta)$ gates, $(k-1)$ controlled-$U_x(2\theta)$ gates, and two instances of controlled-$U_x(\theta)$ gates, all controlled on the mth instance of the output qubit.

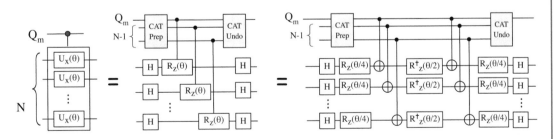

Figure 9.3: The decomposition of the controlled unitary operation $U_x(\theta)$ into single-qubit R_z gates and CNOT gates.

The quantum circuits for the decomposition of the controlled-$U_x(2\theta)$ and controlled-$U_{zz}(2\theta)$ gates are shown in Figures 9.3 and 9.4, respectively. The gates are decomposed into rotations about

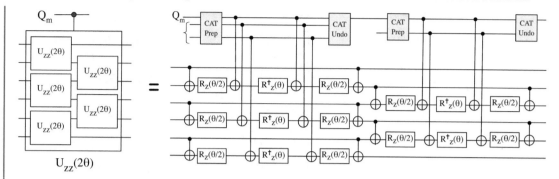

Figure 9.4: The decomposition of the controlled unitary operation $U_{zz}(2\theta)$ gate into single-qubit R_z gates and CNOT gates.

the $\hat{z}$-axis, $R_z(\theta) = \exp(-i\frac{\theta}{2}Z)$ and CNOT gates. $(N-1)$ additional qubits are used to prepare an N-qubit cat state in order to parallelize each of the N $R_z(\theta)$ gates. The preparation of an N-qubit cat state requires $(N-1)$ CNOT gates, which can be implemented in $\mathcal{O}(N)$ time steps in parallel with the $R_z(\theta/4)$ gates in Figure 9.3 and in parallel with the $R_z(\theta/2)$ gates in Figure 9.4.

The three single-qubit R_z gates ($R_z(\theta)$, $R_z(\theta/2)$, and $R_z(\theta/4)$) can be approximated using $\mathcal{O}(\log^{3.97}(1/\epsilon_{sk}))$ basic gates (H, T, S) by the Solovay-Kitaev theorem [61, 117]. The Solovay-Kitaev error (ϵ_{sk}) is equivalent to a small rotation applied to the qubit. The algorithm of Dawson and Nielsen [61] is used to compute the sequence of H, T, and S gates required to approximate each of the three R_z gates for $\theta = \frac{2^m\tau}{k}$. Clark et al. [54] define S_R as the length of the longest of these three sequences. For M=30, for example, we find that $S_R = 4 \times 10^5$, requiring a sixth order Solovay-Kitaev approximation [61]. The results of this calculation show that the Solovay-Kitaev error $\epsilon_{sk} < \frac{\epsilon_T}{k}$, in order that the total error, ϵ_T is less than the required precision $(1/2^{M-m})$, when the authors approximate $U(2^m\tau)$. As a result S_R scales as $\mathcal{O}(\log^{3.97}(k/\epsilon_T)) = \mathcal{O}(M^{3.97})$.

We now have a complete decomposition of the controlled-$U(2^m\tau)$ into the basic gate set $\{X, Z, H, T, S, CNOT, MEASURE\}$. As a function of S_R, the number of time steps required to implement controlled-$U_x(\theta)$ and $U_{zz}(\theta)$ is equal to $(3S_R + 4)$, and $(6S_R + 7)$, respectively. Following Equation (9.7), the number of time steps required to implement the entire controlled-$U(2^m\tau)$ is $k(9S_R + 11) + 3S_R + 4$, where $k = 2^m k_0$. Each R_j gate in Figure 9.1 is equivalent to a rotation by $R_z(\theta)$ and requires less than S_R gates.

Putting all of the above together, the total number of time steps (K) required to implement the TIM circuit as a function of S_R, k_0, and M is given by:

$$K = \sum_{m=0}^{M-1}[2^m k_0(9S_R + 11) + 3S_R + 4 + S_R]$$
$$= \mathcal{O}(2^M) \times S_R. \tag{9.8}$$

Since S_R scales as $\mathcal{O}(M^{3.97})$, the total number of time steps is dominated by the exponential dependence on the precision (M). The number of qubits Q required to implement the circuit is $2N$, since N qubits are needed for the input register $|\Psi\rangle$, one qubit is needed for the output register, and $N - 1$ qubits are needed for the cat state.

Next, the authors include fault-tolerant QEC into the TIM circuit and determine the resulting resource requirements, K and Q when implemented using the QLA architecture. An estimate is also provided on how long it could take to implement the TIM problem in real-time by taking into account the underlying physical implementation of each gate and qubit in the context of the QLA architecture.

9.3   MAPPING THE TIM CIRCUIT ONTO THE QLA ARCHITECTURE

The number of logical qubits Q for the TIM problem directly maps to the number of computational tiles required by the QLA, allowing us to estimate the size of the physical system required for quantum simulation. Similarly, the number of time steps K maps directly to the time required to implement the application since the duration of a single time step in the QLA architecture is defined as the time required to perform error correction.

The parameters K and Q for the TIM problem were estimated in Section 9.2.2, where Q was found to be $2N$ and K is $\mathcal{O}(2^M) \times S_R$. The fault-tolerant implementation of the T gate, however, requires an auxiliary logical qubit prepared into the state $T|+\rangle$ for one time step followed by four time steps composed of H, CNOT, S, and measurement gates [7], causing the value of K and Q to increase. Since many of the gates in the Solovay-Kitaev sequences approximating the R_z gates are T gates, when calculating K using Equation (9.8), the value of S_R must take into consideration the increased number of cycles for each T gate. All other basic gates are implemented transversally and require only one time step.

The resulting functional layout for the QLA architecture for the TIM problem is shown in Figure 9.5. The architecture consists of $4N$ logical qubit tiles. The tiles labeled with Q_1 through Q_N are the data tiles which hold the logical qubits used in the N-qubit input register $|\Psi\rangle$ and the "OUT" tile is for the output register. The tiles labeled with C_1 through C_{N-1} are the $N - 1$ qubit tiles for the cat state. The $T|+\rangle$ tiles are for the preparation of the auxiliary states in the event that T gates are applied on any of the data qubits. All tiles are specifically arranged as shown in Figure 9.5 in order to minimize the communication required for each logical CNOT gate between the control and target qubits. For example, when preparing the cat state using all C_i tiles and the "OUT" tile, CNOT gates are required only between the "OUT" tile, C_1, and C_r. Similarly, C_1 interacts via a CNOT gate only with C_2, while C_2 interacts only with Q_3, during the cat state preparation.

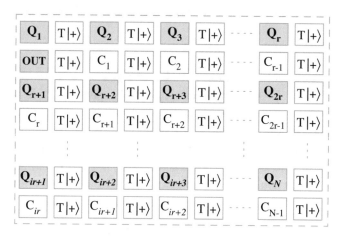

Figure 9.5: QLA architecture for the TIM problem

9.3.1   RESOURCE ESTIMATES FOR THE 1-D TIM PROBLEM

The resource requirements for implementing the 1-D TIM problem using the QLA architecture are given in Figure 9.6, which shows a logarithmic plot of the number of time steps K (calculated using Equation (9.8)) as a function of the energy precision $M \le 20$, assuming $N = 100$. The figure clearly shows K's exponential dependence on M. The dependence of K on the number of spins (N) is negligible and appears only in the k_0 term in Equation (9.8) as $\mathcal{O}(1/N)$, as discussed in Section 9.2.2. In fact, since $Q = 4N$, very little increase is expected in the value of the total problem size KQ as N increases.

It can be seen that for $M \le 8$, no error correction is required. This is because the required reliability per gate of $1/KQ$ is still below the physical ion-trap gate reliability of 1×10^{-7}. Without error correction, the architecture is composed entirely of physical qubits and all gates are physical gates. This means that each single-qubit R_z gate can be implemented directly without the need to approximate it using the Solvay-Kitaev theorem, resulting in $S_R = 1$ in Equation (9.8), and the total number of qubits becomes $2N$ instead of $4N$. For $M \ge 9$ error correction is required, resulting in a sudden jump in the number of timesteps at $M = 9$, with an additional scaling factor of $\mathcal{O}(M^4)$ in K due to S_R's dependence on M. In fact, K increases so quickly that at $M = 9$ that level 2 error correction is required instead of level 1. At $M \ge 18$ level 3 error correction is required and, while there is no increase in K, each time step is much longer, so there is a jump in the number of days of computation. The Solovay-Kitaev order [61] for $M = 9$ is three and increases to order five for $M = 20$.

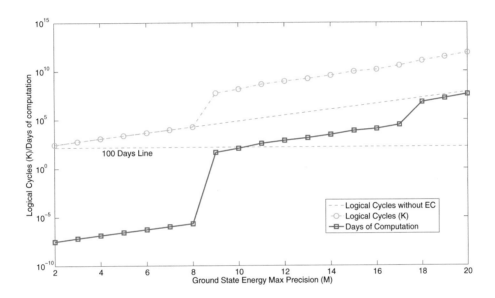

Figure 9.6: (color online) Numerical calculations for the number of logical cycles K (solid line) and days of computation necessary, assuming $N = 100$ spin TIM problem as a function of the desired maximum precision $M \leq 20$.

CHAPTER 10

Teleportation-Based Quantum Architectures

It would be ideal to treat the physical implementation of quantum logic gates with a specific technology in mind such as the QLA's treatment of ion-traps. The problem is that there is an enormous amount of available choices for physical gate implementation, which is equally matched by an enormous amount of available possibilities for constructing logically universal circuits. In this chapter, we consider architectures in which we take advantage of the teleportation as a computational primitive [87] to implement logic gates through teleportation circuits that are more suitable to other quantum computation technologies.

In Section 2, we described the circuit model for quantum computation which implements universal quantum logic as a sequence of unitary matrices that act on the probability amplitude vector describing a collection of units of quantum data known as qubits. Furthermore, in Section 2, we introduced the Clifford group operations (see Equation 4.10) combined with the single-qubit T gate as an elementary basis for universal quantum computation. The chosen basis of gates offers relatively straight forward, fault-tolerant constructions for implementing quantum logic on encoded qubits. Steane [197] summarizes several proposals for constructing a fault-tolerant universal set of quantum operations that includes the Clifford group. Some proposals include the three-qubit Toffoli gate as an elementary operation, and some the controlled-S gate [124, 186]. When designing a quantum architecture, or modeling software for quantum architectures, a system designer may need to allow flexibility in the software to choose the appropriate universal set of gates that allows the generalization to all $[[n, k, d]]$ error correcting codes.

Perhaps, even more interesting is the fact that we may not even need the direct application of gates to perform universal quantum computation. All we need is a circuit structure (or a mechanism) that implements the functionality of universal gates. More specifically, a mechanism that allows the unitary transformation of a quantum state $|\Psi\rangle$ without the physical application of the unitary operation itself. Such a mechanism is *teleportation*. In 1998 Gottesman and Chuang published a paper [87] that showed teleportation as a universal quantum logic primitive that can be used to perform any quantum computation. The teleportation gate scheme is used to allow two-qubit operations in optical quantum computers (see Section 4.1.1). We can extend this further by looking at the possible tradeoffs when designing an architecture that utilizes universal quantum logic on encoded data through teleportation. Such investigations may drastically change the structure of the entire quantum system as defined by the QLA architecture case study.

One of DiVincenzo's principal requirements for quantum technologies is the ability to orchestrate universal quantum logic, which is generally composed of single-qubit gates, two-qubit gates, and measurement in the circuit model of quantum computation. Save for superconducting qubits, most technologies allow for relatively easy arbitrary single-qubit rotations. Therefore, the ability to perform qubit-qubit interactions, or two-qubit gates is the most critical requirement for a given technology. Particularly, since an implicit assumption in any qubit-qubit interaction is the ability to communicate quantum information between the two qubits.

Many of the circuit synthesis papers mentioned in Section 8.3 assume the CNOT gate to be the standard elementary two-qubit gate and synthesize circuits to be CNOT-optimal. Perhaps incorrectly, a general assumption is that DiVincenzo's criteria demands the ability for a technology to demonstrate a reliable CNOT gate; however, the direct application of a CNOT gate is not necessarily required. The elementary two-qubit gate in the ion-trap technology, for example, is the controlled-Z rotation [18, 171], which can be used to functionally construct a CNOT gate as shown in Figure 10.1.

Figure 10.1: A CNOT gate can be built by using a controlled-Z gate and two Hadamard gates.

Any two-qubit gate used to implement a CNOT operation requires the interaction of two qubits, either through *direct* qubit-qubit interaction, which implies that the quantum data for both qubits is placed the same spatial location through teleportation or through direct physical movement of the qubit carriers; or *indirect* qubit-qubit interaction, which is done through some shared medium that allows the two-qubit states to be coupled without the need to bring the data spatially close together.

Both types of interactions have their respective drawbacks. Qubits that interact directly require either information swapping between nearest neighbors, or shuttling qubits through empty channels: introducing errors proportional to the length of the channels. Transferring quantum information creates the need for complex low-level schedulers such as QPOS [141], or the inner-workings of QUALE [14], both of which assume technologies that require direct physical qubit communication. In addition, bringing the states of two qubits together creates difficulties for distinguishing the qubits from one another and opportunities for correlated errors.

The indirect qubit-qubit interaction may seem more efficient on the outset by leaving the qubits in place, but it still requires a common medium that is used to couple the qubits and can potentially introduce correlated errors. There are several techniques to achieve this: One technique uses single photons to implement multi-qubit gates between trapped atoms [31, 71, 145], another technique couples qubits through a common quantum field mode, which can be thought of as a shared quantum "bus" and can be realized with a laser beam [138, 192].

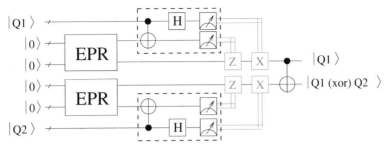

Figure 10.2: Standard CNOT operation between two logical qubits in remote locations. The qubits are teleported to a common destination such as two adjacent accumulators in a processing element and interacted with a CNOT gate.

10.1   THE CNOT GATE AND SINGLE-QUBIT GATES THROUGH TELEPORTATION

The simplest way to consider the implementation of a CNOT gate using teleportation is already utilized by the QLA architecture we described in Section 7. A schematic is shown in Figure 10.2. Qubits Q_1 and Q_2 residing in the memory region are teleported to a processing element (PE) through EPR pairs created between each respective memory address and the PE.

The gates in the dashed boxes in the circuit of Figure 10.2 implement a *Bell Measurement* between any two qubits. Recall the four two-qubit Bell states $\{|\Psi_+\rangle, |\Psi_-\rangle, |\Phi_+\rangle, |\Phi_-\rangle\}$ given in Equation 7.2, where the state $|\Psi_+\rangle$ is the familiar two-qubit EPR state. Just as a single-qubit state can be written as a superposition of two basis states such as $|+\rangle$ and $|-\rangle$ or $|0\rangle$ and $|1\rangle$, a two-qubit state can be written as a superposition of the four Bell states:

$$|q_1, q_2\rangle = c_0|\Psi_+\rangle + c_1|\Psi_-\rangle + c_2|\Phi_+\rangle + c_3|\Phi_-\rangle, \tag{10.1}$$

A Bell measurement such as the circuit in the dashed boxes of Figure 10.2 determines which of the four Bell states the two qubits are in. As a result of the measurement, the two qubits are collapsed into one of the Bell states. The Bell measurement also serves as an entangling operation between the two qubits if they are originally unentangled. We will abstract the Bell measurement procedure as a box much like we did with the EPR pair creation because each technology has a very specific method for implementing a Bell measurement on two-qubits:

Figure 10.3: We abstract the Bell measurement circuit as a box with the inscription "Bell".

In Section 7, we described the long-distance communication channel as a repeater-based interconnect which creates 49 purified elementary EPR pairs that span the entire channel between qubits Q_1 and Q_2. The teleportation procedure transfers their 49-qubit encoded states onto the 49 elementary EPR pair qubits at the two destination accumulators. The direct application of a transversal CNOT gate follows the logical qubit transfer once they are located in adjacent accumulators.

To avoid the direct interaction between the two logical qubits, we can move the CNOT gate through the single-qubit X and Z operations of Figure 10.2 by changing their order without affecting the functionality of the circuit. The result is shown in Figure 10.4. There is no direct interaction between qubits Q_1 and Q_2, but there is a direct CNOT gate between the EPR blocks. The interaction between the EPR blocks is only possible if the four blocks themselves are encoded logical qubits initially in the logical $|\overline{0}\rangle$ states as shown in the figure. In this manner, the implementation of the CNOT gate decomposes into encoding 4 qubits initialized to $|\overline{0}\rangle$ into some pre-specified 4-qubit state, denoted as the state $|\overline{M}\rangle$, where:

$$|\overline{M}\rangle = \frac{(|\overline{00}\rangle + |\overline{11}\rangle)|\overline{00}\rangle + (|\overline{01}\rangle + |\overline{10}\rangle)|\overline{11}\rangle}{\sqrt{2}} \tag{10.2}$$

Moreover, the $|\overline{M}\rangle$ state can be prepared using two 3-qubit cat-states ($|000\rangle + |111\rangle$), also known as GHZ states [88], by applying a Bell measurement between two of the qubits in each GHZ state followed by single-qubit gates controlled on the result of the Bell measurement [87]. The 4 qubits not involved in the Bell measurement will retain the $|\overline{M}\rangle$ state and can be used for the implementation of the CNOT gate. This gives us a CNOT gate mechanism that requires only classically controlled single-qubit gates, a specially created 4-qubit entangled state, and two Bell basis measurements. Given that the creation of the $|\overline{M}\rangle$ state can be performed offline, and deterministically much like the creation of EPR qubits, then the CNOT gate between two logical qubit may be implemented without any direct qubit-qubit interaction.

Remotely entangling two qubits to form an EPR pair is possible [31, 71, 145]. In addition, it may be possible to remotely entangle three qubits into a GHZ state, or even create a black-box mechanism that creates GHZ states of encoded qubits and distributes them in the architecture through a repeater-based channel as used in the QLA model. Even if the black-box consists of traditional data shuttling to create qubit-qubit interactions for the encoded special states, it can be localized to a special state "factory" region where the distances are short relative to the application level system.

Gottesman and Chuang [87] further show that the same methodology can be used to construct a teleportation-based mechanism for any encoded single-qubit logical operation. A schematic for implementing an arbitrary single-qubit operator $\overline{U}$ is shown in Figure 10.5.

The universal gate-set we are considering only requires a teleportation implementation for the T gate for any other single-qubit gate is transversal and can be applied locally. The T gate circuit shown in Figure 4.9 of Section 4.2.3 is much simpler than the network of Figure 10.5 and utilizes the concept of *one-bit teleportation* [229]; however, it requires a CNOT gate between the data state and the specially prepared $|A_{\pi/8}\rangle$ ancilla state. To avoid direct qubit-qubit interaction, the required

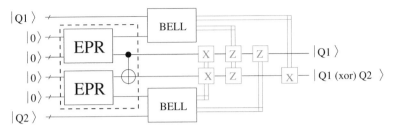

Figure 10.4: Teleporting two-qubits through a controlled-not gate by using only single-qubit rotations, Bell measurements, and a special 4-qubit state $|\overline{M}\rangle$ which can be composed of two EPR pairs, or two GHZ states. If two GHZ states are used, the CNOT gate between the two EPR pairs will be replaced by a Bell measurement between the two GHZ states. The circuit shown and the procedure is given in Reference [87].

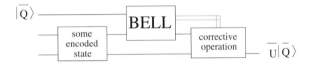

Figure 10.5: An implementation of a single-qubit operator U through teleportation. The implementation requires the preparation of an encoded state using two logical ancillary qubits and a Bell measurement followed by the corrective operation.

CNOT gate in Figure 4.9 can be implemented using the teleportation circuit shown in Figure 10.2 with the resource cost of four additional qubits for the creation of the $|M\rangle$ state.

10.2   THE ARCHITECTURE

We are faced with two gate implementation choices. The first one is to teleport data to a processing region using the repeater-based interconnect, and the second choice is to teleport gates through specially created ancillary states. Figure 10.6 shows the distinction between the two choices of distributing quantum computation in the two-bit adder from Section 2.4. The adder is divided into two main processing regions that initially perform computation in parallel through the first two timesteps. The third timestep requires a Toffoli gate between the ancilla qubit in the first region and two qubits from the lower (second) region. The Toffoli gate has been decomposed into elementary one and two-qubit gates in the dashed box of each half of Figure 10.6. If the two 3-qubit regions are significantly far apart, we have a choice to teleport the data as described in Section 7.3, or to teleport the qubits through the gates involved in the decomposition of the Toffoli gate (see Figure 2.4 in Section 2.4).

Reference [213] discusses the tradeoffs associated with teleporting gates as we discussed so far, and teleporting data on a distributed quantum computer, where the schematic distinction shown

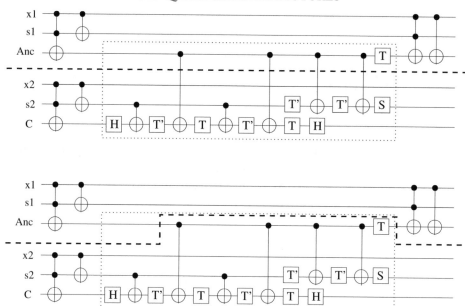

Figure 10.6: The model used by Reference [213] to distinguish between teleporting data and teleporting gates in a distributed quantum computer. The north half of the figure shows a 2-bit adder of 6 qubits where the middle Toffoli gate has been expanded into its one and two-qubit gate decomposition. The think dashed line separates the two processing regions evenly as it is intended to illustrate that gates are teleported from one region to the other, while the data remains in place. The south part of the figure shows data teleportation as described by the QLA architecture.

in Figure 10.6 is introduced. The authors base their study on several implementations of the adders used for Shor's factoring algorithm and find that, at the large-scale, it is more expensive to teleport gates than it is to teleport data in terms of the number of elementary operations competing for shared resources. The authors of Reference [213] use a clever construction of the teleported CNOT gate that does not require the four ancillary qubits to be placed in the $|\overline{M}\rangle$ state and leaves the CNOT gate implementation in the encoded data qubits, rather than in the EPR qubits. Their construction is shown in Figure 10.7.

The underlying architecture is based on solid-state qubits coupled indirectly through a universal quantum bus [138, 192]. Similar distributed architecture can be realized with the ion-trap technology by coupling two ions through photo detector stations and beam splitters [31, 71, 145]. The quantum bus connects the distributed pieces of the application level system, where each piece uses transceiver qubits to connect to the bus. In this manner, data or gates can be transferred be-

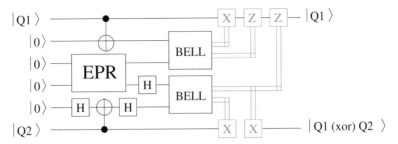

Figure 10.7: Teleporting two-qubits through a controlled-not gate that requires 4 ancillary qubits, but only 2 need to be encoded as an EPR pair. The other two are used to couple with the data before the Bell measurement operation.

tween multiple distributed regions by using the transceiver qubits as EPR pairs for teleportation. Two transceiver qubits in different regions can be remotely entangled through the quantum bus.

Intuitively, the observation that teleporting gates is less efficient than teleporting data is reasonable when looking at Figure 10.6. Once the data is teleported to a specific region, it becomes *local* to that region and the sequence of gates can be executed directly to complete the Toffoli operation without much communication overhead. On the other hand, the teleportation of gates requires repeated usage of the quantum bus and the contention for the transceiver qubits increases [213].

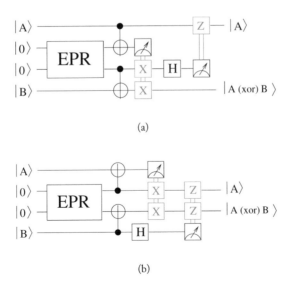

Figure 10.8: Two versions of a simplified "remote" CNOT gate.

But what about encoded gates on fault-tolerantly constructed logical qubit states, which will undoubtedly be required for large-scale applications? The CNOT gate construction in Figure 10.7 is not a truly teleported CNOT gate because it requires two local CNOT operations between the data qubits and the ancillary qubits before the Bell measurement procedure. If locally executed CNOT gates are allowed where data is transferred between the control qubit and the target qubit, then we can use much simpler teleportation-based CNOT gate construction given in Reference [229]. Figure 10.8 shows two versions of a teleported CNOT gate where only two ancillary qubits are required as an encoded EPR pair between the control and target qubits.

An interesting tradeoff would be to study the optimal logical distances at which traditional direct interaction CNOT gates are allowed, and larger distances where CNOT gates are sent through teleportation procedures through remote data coupling. Our architecture can be a distributed logical architecture, where there are N regions labeled $\{R_1, R_2, ...R_N\}$ which contain both logical data qubits and logical ancillary qubits used for the creation of specialized states for gate teleportation. This has the potential to significantly improve the reliability of the application. Standard direct interaction logical CNOT gates are executed within each region. The logical data never leaves to another region, but inter-region CNOT gates are implemented through the specialized ancillary qubits in each region.

This scheme has the potential to significantly improve the reliability of the architecture, as logical gate distances between regions are relatively short, and inter-region gates are teleported. The specialized states between regions can be prepared independently of the execution of the application and verified for correctness. The coupling of the individual qubits can be done remotely through entangling trapped-ions through fiberoptic wires or using the shared quantum bus. Only specialized states that pass the verification procedures will be used for gate teleportation where Bell measurements are performed. Of course, this scheme would require sufficient amount of resources invested in the preparation of the specialized ancillary states for gate teleportation. The logical data qubits and specialized ancillary qubits would necessarily be equipped with the error correction mechanisms needed for each logical qubit tile, further increasing the amount of error correction resources.

An alternative construction would be to use gate teleportation to speed up quantum applications. For example, if qubits Q_1 and Q_2 are required for a certain sequence of operations and the two qubits reside in the memory region, the first operation in the program can be performed while teleporting the qubits to the processing region.

10.3   ERROR CORRECTION THROUGH TELEPORTATION

Even more amazing, are the possible error correction advantages gained when allowing sufficient interconnect bandwidth such that *encoded* EPR pairs are communicated instead of elementary EPR pairs as in the QLA architecture. Notice, the relationship between the control data qubit and the nearest EPR qubit in Figure 10.8(a) (lines 1 and 2). If the qubits are encoded using some CSS $[[n, k, d]]$ code such as the Steane $[[7, 1, 3]]$ code, then the sequence of operations between lines 1 and 2 is strikingly similar to the Steane method for error correction. As a matter of fact it is, and while we are teleporting the gate, the measurement performed is equivalent to extracting the error syndrome of the data.

In fact, teleportation itself is error correction. Let's take a step back and consider the standard teleportation procedure that simply teleports quantum data as in the original circuit of Figure 2.7 in Section 2.6, or the CNOT gate from Figure 10.2. If less than $t = (d - 1)/2$ errors have occurred on the logical data by the time the Bell measurement is complete, then the encoded state of the qubit will be *correctly* identified through the logical measurement operation. The correct state will then be recreated at the destination EPR logical qubit. If the EPR qubit is sufficiently well distilled, teleportation is another method for correcting errors on encoded data.

Knill [119, 120, 123] has studied the fault-tolerance of using teleportation as error correction protocol applied on linear optical architectures. He has devised extensive *error detecting code* procedures and has demonstrated that the accuracy threshold for scalable quantum computation can be as high as 1% error rate per physical gate. His method of *postselected quantum computation* uses the property that logical states used for computation are accepted only if no errors are detected with sufficiently high probability. He uses simple four and six-qubit concatenated quantum error detecting codes to show that, by postselecting the output of the logical operations, the probability of error in his architecture can be reduced arbitrarily. All quantum logic is performed through teleportation of gates.

CHAPTER 11

Concluding Remarks

In this book, we have explored the design of large-scale, quantum architectures in the context of system-level balance between fault-tolerant, logical qubit structures and communication mechanisms that protect quantum data while in transmission. Logical qubits structures include the number of ancillary qubits necessary for the required rate of error correction. The bandwidth of the interconnect channels is balanced with the size and speed of the computational blocks that work on these logical qubits. The distribution of the quantum computational resources is matched to the application's support for gate teleportation or data teleportation, and thus allowing for the creation of logical teleportation resources. The amount of usage for different error correcting codes or levels of encoding is matched to the size of the application and the needed reliability to finish the application with a high enough success rate. In general, the inherently high decoherence rate of quantum information places the issue of fault-tolerance at the heart of a balanced system design.

Design of large-scale quantum systems is in its infancy. As quantum technologies continue to improve, however, the opportunities for system designers will dramatically increase. There are already several groups exploiting these opportunities:

- Emanuel Knill at the National Institute for Standards and Technology (NIST) is the leading architect behind fault-tolerant optical systems with teleportation based error correction and gate implementations [120, 123].

- Mark Oskin and David Bacon at the University of Washington are working to design tools to study and model quantum architectures based on some of the most efficient error correcting codes known [11, 153, 154, 164, 216].

- Mircea Vladutiu from the Politehnica University of Timisoara, Romania has published extensively about modeling quantum algorithms on reconfigurable circuit structures that use reconfigurability to improve the scalability of quantum error correcting codes [207, 208]

- Researchers at the Quantum Architectures Research Center (QARC) led by John Kubiatowicz at UC Berkeley, Mark Oskin at the University of Washington, Isaac Chuang at MIT, and Frederic T. Chong at UC Santa Barbara have made a significant impact on the studying of the implementation and control of classical control structures for emerging quantum technologies [101, 102, 103, 154, 220].

- Teleportation-based distributed quantum systems for large-scale applications are being studied at Keio University, Japan guided by Kohei Itoh [213, 216].

- The quantum circuits group from the University of Michigan, together with collaborators at MIT and Columbia, have done extensive contributions to quantum logic circuit synthesis and testing, including the development of fault-tolerant software architecture for quantum computers that maps a high-level program into fault-tolerant machine-level instructions [159, 181, 183, 199, 200].

- Other groups not listed here, combined with the numerous theoretical and experimental research projects that are ongoing, make the field of quantum computing one of the fastest growing fields of science.

In this book, we have focused on the QLA architecture as a case study from which to develop a framework of architectural abstractions. To model the QLA architecture we have made some very strict design assumptions such as the fault-tolerant structure of the long-distance interconnect, the error correcting code of the encoded qubits, and, finally, the low-level microarchitecture model based on the ion-trap technology. While the assumptions made are sufficient to demonstrate that, within existing technological boundaries, scalable quantum computation is feasible– there are still many possibilities for constructing the basic fault-tolerant elements of an architecture. Our hope is that this book will help provide the necessary background and abstractions for system designers to explore this space of technologies and potential designs. Leveraging our collective experience in computer design will be instrumental in making practical quantum computing a reality.

Bibliography

[1] S. Aaronson and D. Gottesman. Improved simulation of stabilizer circuits. *Phys. Rev. A*, 70:052328, 2004, quant-ph/0406196. DOI: 10.1103/PhysRevA.70.052328 90, 134

[2] D. Abrams and S. Lloyd. Simulation of many-body fermi systems on a universal quantum computer. *Phys. Rev. Lett.*, 79:2586–2586, 1997. DOI: 10.1103/PhysRevLett.79.2586 149, 150, 151

[3] D. Abrams and S. Lloyd. Quantum algorithm providing exponential speed increase for finding eigenvalues and eigenvectors. *Phys. Rev. Lett.*, 83:5162–5162, 1999. DOI: 10.1103/PhysRevLett.83.5162 149, 150, 151

[4] D. Aharonov and M. Ben-Or. Fault tolerant computation with constant error. pages 176–188, quant-ph/9906129. 57, 72, 130

[5] D. Aharonov, W. van Dam, J. Kempe, Z. Landau, S. Lloyd, and O. Regev. Adiabatic quantum computation is equivalent to standard quantum computation. 2004. DOI: 10.1109/FOCS.2004.8 21

[6] P. Aliferis, F. Brito, D.P. DiVincenzo, J. Preskill, M. Steffen, and B.M. Terhal. Fault-tolerant computing with biased-noise superconducting qubits: a case study. *New Journal of Physics*, 11:013061, 2009. DOI: 10.1088/1367-2630/11/1/013061 72

[7] P. Aliferis, D. Gottesman, and J. Preskill. Quantum accuracy threshold for concatenated distance-3 codes. *Quant. Inf. Comp.*, 6:97–165, 2007. 90, 94, 120, 131, 139, 142, 155

[8] S. Amash and et al. Toward the manipulation of a single spin in an algaas/gaas single-electron transistor. *Proceedings of the SPIE Defense and Security Symposium, Orlando FL*, 2006. DOI: 10.1117/12.665548 52

[9] A. Ambainis and O. Regev. An elementary proof of the quantum adiabatic theorem. *ArXiv Quantum Physics e-prints*, 2004, quant-ph/0411152v2. 46

[10] A. Aspuru-Guzik, A. Dutoi, P. Love, and M. Head-Gordon. Simulated quantum computation of molecular energies. *Science*, 309:1704–1704, 2005. DOI: 10.1126/science.1113479 151

[11] D. Bacon. Operator quantum error correcting subsystems for self-correcting quantum memories. 2005, quant-ph/0506023. DOI: 10.1103/PhysRevA.73.012340 94, 131, 169

[12] F. Bahr, M. Boehm, J. Franke, and T. Kleinjung. Rsa-640 is factored! (Online Arxive) `http://www.rsasecurity.com/rsalabs/node.asp?id=2964`, 2005. 77

[13] S. Balensiefer, L. Kregor-Stickles, and M. Oskin. An evaluation framework and instruction set architecture for ion-trap based quantum micro-architectures. *in Proc. ISCA-32; Madison, WI*, 2005. DOI: 10.1145/1080695.1069986 88, 104, 130

[14] S. Balensiefer, L. Kregor-Stickles, and M. Oskin. Quantum architecture tools: Quale. *Available for download online at: http://www.cs.washington.edu/homes/lucasks/tools.html.* DOI: 10.1117/12.604073 88, 130, 131, 160

[15] S. Balensiefer, L. Kregor-Stickles, and M. Oskin. An evaluation framework and instruction set architecture for ion-trap based quantum micro-architectures. *in Proc. ISCA-32; Madison, WI*, 2005. DOI: 10.1145/1080695.1069986 138

[16] A. Barenco, C.H. Bennett, R. Cleve, D.P. DiVincenzo, N. Margolus, P. Shor, T. Sleator, J. Smolin, and H. Weinfurter. Elementary gates for quantum computation. *Phys. Rev. A.*, 52:3457, 1995, quant-ph/9503016. DOI: 10.1103/PhysRevA.52.3457 134

[17] S.M. Barnett and S.J.D. Phoenix. Information-theoretic limits to quantum cryptography. *Phys. Rev. A*, 48(1):R5–R8, 1993. DOI: 10.1103/PhysRevA.48.R5 3

[18] M. Barrett, J. Chiaverini, T. Schaetz, J. Britton, and et. al. Deterministic quantum teleportation of atomic qubits. *Nature*, 429, 2004. DOI: 10.1038/nature02608 51, 52, 54, 79, 108, 141, 160

[19] J.S. Bell. On the einstein-podolsky-rosen paradox. *Physics*, 1:195–200, 1964. 110

[20] P. Benioff. Quantum mechanical models of turing machines that dissipate no energy. *Phys. Rev. Lett.*, 48:15811585, 1982. DOI: 10.1103/PhysRevLett.48.1581 3

[21] C.H. Bennett and et. al. Teleporting an unknown quantum state via dual classical and EPR channels. *Phys. Rev. Lett.*, 70:1895–1899, 1993. DOI: 10.1103/PhysRevLett.70.1895 4, 17, 79

[22] C.H. Bennett and et. al. Purification of noisy entanglement and faithful teleportation via noisy channels. *Phys. Rev. Lett.*, 76:722, 1996. DOI: 10.1103/PhysRevLett.76.722 81, 109

[23] C.H. Bennett, F. Bessette, G. Brassard, L. Salvail, and J. Smolin. Experimental quantum cryptography. *Journal of Cryptography*, 5(1), 1992. DOI: 10.1007/BF00191318 3

[24] C.H. Bennett and G. Brassard. Quantum cryptography: Public key distribution and coin tossing. *IEEE International Conference on Computers, Systems, and Signal Processing*, pages 175–179, 1984. 3, 53

[25] C. Bennett and S.J. Wiesner. Communication via one- and two-particle operators on einstein-podolsky-rosen states. *Phys. Rev. Lett.*, 69(2881), 1992. DOI: 10.1103/PhysRevLett.69.2881 19

[26] E. Bernstein and U. Vazirani. Quantum complexity theory. *SIAM Journal on Computing*, 26(5):1411 – 1473, 1997. DOI: 10.1145/167088.167097 3

[27] J.D. Biamonte and P.J. Love. Realizable hamiltonians for universal adiabatic quantum computers. *Phys. Rev. A*, 78(7):012352, 2008. DOI: 10.1103/PhysRevA.78.012352 151

[28] A. Blais, R-S. Huang, A. Wallraff, S.M. Girvin, and R.J. Schoelkopf. Cavity quantum electrodynamics for superconducting electrical circuits: An architecture for quantum computation. *Phys. Rev. A*, 69(062320), 2004. DOI: 10.1103/PhysRevA.69.062320 51

[29] B.B. Blinov, L. Deslauriers, P. Lee, M.J. Madsen, R. Miller, and C. Monroe. Sympathetic cooling of trapped cd+ isotopes. *Phys. Rev. A.*, 65:040304, 2002. DOI: 10.1103/PhysRevA.65.040304 52, 54

[30] B.B. Blinov, L. Deslauriers, P. Lee, M.J. Madsen, R. Miller, and C. Monroe. Sympathetic cooling of trapped ions for quantum logic. *Phys. Rev. A.*, 61:032310, 2000, quant-ph/9909035. DOI: 10.1103/PhysRevA.61.032310 54, 56

[31] B.B. Blinov, D.L. Moehring, L.M. Duan, and C. Monroe. Observation of entanglement between a single trapped atom and a single photon. *Nature*, 428:153–157, 2004. DOI: 10.1038/nature02377 53, 160, 162, 164

[32] M. Bocko, A. Herr, and M. Feldman. Prospects for quantum coherent computation using superconducting electronics. *IEEE Trans. App. Supercond*, 7:3638–3641, 1997. DOI: 10.1109/77.622206 52

[33] P. Bonderson, A. Kitaev, and K. Shtengel. Detecting non-abelian statistics in the v=5/2 fractional quantum hall state. *Phys. Rev. Lett.*, 96(016803), 2006. DOI: 10.1103/PhysRevLett.96.016803 21

[34] P. Bonderson, K. Shtengel and J.K. Slingerland. Decoherence of anyonic charge in interferometry measurements. *Phys. Rev. Lett.*, 98(070401), 2007. DOI: 10.1103/PhysRevLett.98.070401 21

[35] D. Boschi, S. Branca, F. De Martini, L. Hardy, and S. Popescu. Experimental realization of teleporting an unknown pure quantum state via dual classical and einstein-podolsky-rosen channels. *Phys. Rev. Lett.*, 80:11211125, 1998. DOI: 10.1103/PhysRevLett.80.1121 53

[36] D. Bouwmeester and et. al. Experimental quantum teleportation. *Nature*, 390:575–579, 1997. DOI: 10.1038/37539 53, 79

[37] H.J. Briegel and R. Raussendorf. Persistent entanglement in arrays of interacting particles. *Phys. Rev. Lett*, 86:910913, 2001, quant-ph/0004051. DOI: 10.1103/PhysRevLett.86.910 21, 54

[38] J. Britton, D. Leibfried, J.Beall, R.B. Blakestad, J.J. Bollinger, J. Chiaverini, R.J. Epstein, J.D. Jost, D. Kielpinski, C. Langer, R. Ozeri, R. Reichle, S. Seidelin, N. Shiga, J.H. Wesenberg, and D.J. Wineland. A microfabricated surface-electrode ion trap in silicon. 2006, quant-ph/0605170. 52, 120

[39] K.R. Brown, R.J. Clark, and I.L. Chuang. Limitations of quantum simulation examined by simulating a pairing hamiltonian using nuclear magnetic resonance. *Physical Review Letters*, 97:050504, 2006. 151, 152

[40] J.P. Buhler, H.W. Lenstra, and C. Pomerance. Factoring integers with the number field sieve. *Pages 50-94 in The Development of the Number Field Sieve, volume 1554 of Lecture Notes in Mathematics*, Springer-Verlag, Berlin, 1994. 3, 77

[41] S.S. Bullock and I.L. Markov. Asymptotically optimal circuits for arbitrary n-qubit diagonal computations. *Quantum Information and Computation*, 4(1):027–047, 2004, quant-ph/0303039. 134

[42] C. Cabrillo and et. al. Creation of entangled states of distant atoms by interference. *Phys. Rev. A*, 59:1025–1033, 1999. DOI: 10.1103/PhysRevA.59.1025 52, 80

[43] A.R. Calderbank, E.M. Rains, P.W. Shor, and N.J.A. Sloane. Enlargement of calderbank-shor-steane quantum codes. *IEEE Transcactions of Information Theory*, 45:2492–2495, 1999, quant-ph/9802061. DOI: 10.1109/18.796388 58

[44] A.R. Calderbank and Peter W. Shor. Good quantum error-correcting codes exist. *Physical Review A*, 54:1098, 1996. DOI: 10.1103/PhysRevA.54.1098 58, 66

[45] F.E. Camino, W. Zhou, and V.J. Goldman. Aharonov-bohm electron interferometer in the integer quantum hall regime. *Phys. Rev. B*, 72(155313), 2005. DOI: 10.1103/PhysRevB.72.155313 21

[46] C. Chekuri, R. Johnson, R. Motwani, B. Natarajan, B. Rau, and M. Schlansker. Profile-driven instruction level parallel scheduling with applications to superblocks. *In Proc. 29th Int. Symp. on Microarchitecture*, 29:58–67, 1996. DOI: 10.1109/MICRO.1996.566450 131

[47] J. Chiaverini, R.B. Blakestad J. Britton, J.D. Jost, C. Langer, D. Leibfried, R. Ozeri, and D.J. Wineland. Surface-electrode architecture for ion-trap quantum information processing. *E-Print: quant-ph/0501147*, 2004, quant-ph/0501147. 55

[48] A.M. Childs, E. Farhi, and J. Preskill. Robustness of adiabatic quantum computation. *Phys. Rev. A*, 65, 2002, quant-ph/0108048. 3, 21

[49] H. Chou and C. Chung. An optimal instruction scheduler for superscalar processor. *IEEE Trans. on Parallel and Distributed Systems*, 6(3):303–313, 1995. DOI: 10.1109/71.372778 131

[50] E.M. Chow, H.T. Soh, H.C. Lee, J.D. Adams, S.C. Minne, G. Yaralioglu, A. Atalar, C.F. Quate, and T.W. Kenny. Integration of through-wafer interconnects with a two-dimensional cantilever array. *Sensors and Actuators*, 83:118–123, 2000. DOI: 10.1016/S0924-4247(99)00381-7 56

[51] I.L. Chuang. Quantum algorithm for clock synchronization. *Phys. Rev. Lett.*, 85:2006, 2000, quant-ph/0005092. DOI: 10.1103/PhysRevLett.85.2006 3

[52] I. Cirac and P. Zoller. A scalable quantum computer with ions in an array of microtraps. *Nature*, 404:579âŁ"–581, 2000. DOI: 10.1038/35007021 108

[53] J.I. Cirac and P. Zoller. Quantum computations with cold trapped ions. *Phys. Rev. Lett*, 74:4091–4094, 1995. DOI: 10.1103/PhysRevLett.74.4091 52, 54

[54] C.R. Clark, K.R. Brown, T.S. Metodi, and S.D. Gasster. Resource requirements for fault-tolerant quantum simulation: The transverse ising model ground state. *Phys. Rev. A*, 79(6):062314, 2009. DOI: 10.1103/PhysRevA.79.062314 149, 150, 152, 153, 154

[55] R. Cleve, A. Ekert, C. Macchiavello, and M. Mosca. Quantum algorithms revisited. *Proc. R. Soc. Lon. A*, 454:339–354, 1998. DOI: 10.1098/rspa.1998.0164 151

[56] R. Cleve, A. Ekert, C. Macchiavello, and M. Mosca. Quantum algorithms revisited. *In Proceedings of the Royal Society of London*, A(454):339–354, 1997, quant-ph/9708016. DOI: 10.1098/rspa.1998.0164 17

[57] D. Copsey and et. al. Toward a scalable, silicon-based quantum computing architecture. *Selected Topics, Journal of Quantum Electronics*, 9(6):1552–1569, 2003. DOI: 10.1109/JSTQE.2003.820922 50

[58] D. Cory, A. Fahmy, and T. Havel. Nuclear magnetic resonance spectroscopy: an experimentally accessible paradigm for quantum computing. *In Proceedings of the 4th Workshop on Physics and Computation, New England Complex Systems Institute.*, 1996. DOI: 10.1016/S0167-2789(98)00046-3 52, 80

[59] A. Cross. qasm-tools: An interoperable open-source software tool chain for studying fault-tolerant quantum circuits. *Available for download online at: http://web.mit.edu/awcross/www/qasm-tools/*. 90, 142

[60] D.T. Darshan, T.S. Metodi, A.W. Cross, F.T. Chong, and I.L. Chuang. Quantum memory hierarchies: Efficient designs to match available parallelism in quantum computing. *International Symposium of Computer Architecture (ISCA-33), Boston, MA*, 2006. DOI: 10.1109/ISCA.2006.32 xii, 1, 2, 82, 120, 121, 125

[61] C.M. Dawson and M.A. Nielsen. The solovay-kitaev algorithm. *Quant. Inf. Comp.*, 6:81–95, 2005. 154, 156

[62] B.L. Deitrich and W.M.W. Hwu. Speculative hedge: Regulating compile-time speculation against profile variations. *Proceedings of the 29th International Symposium on Microarchitecture*, 29, 1996. DOI: 10.1109/MICRO.1996.566451 131

[63] D. Deutsch. Quantum computational networks. *Proc. R. Soc. Lond.*, A 400:97–117, 1985. DOI: 10.1098/rspa.1989.0099 3

[64] D. Deutsch. Quantum theory, the church-turing principle and the universal quantum computer. *Proceedings of the Royal Society of London*, A-400:97–117, 1985. DOI: 10.1098/rspa.1985.0070 3, 18

[65] D. Deutsch, A. Ekert, R. Jozsa, C. Macchiavello, S. Popescu, and A. Sanpera. Quantum privacy amplification and the security of quantum cryptography over noisy channels. *Phys. Rev. Lett.*, 77:2818–2821, 1996. DOI: 10.1103/PhysRevLett.77.2818 3, 81

[66] D. Deutsch and R. Jozsa. Rapid solution of problems by quantum computation. *Proceedings of the Royal Society of London*, A-439:553–558, 1992. DOI: 10.1098/rspa.1992.0167 18

[67] D.P. DiVincenzo. The physical implementation of quantum computation. *Fortschr. Phys.*, 48:771–783, 2000, quant-ph/0002077. DOI: 10.1002/1521-3978(200009)48:9/11%3C771::AID-PROP771%3E3.0.CO;2-E 49, 50

[68] T. Draper. Addition on a quantum computer. *ArXiv Quantum Physics e-prints*, 2000, arXiv:quant-ph/0008033. 126

[69] T.G. Draper, S.A. Kutin, E.M. Rains, and K.M. Svore. A logarithmic-depth quantum carry-lookahead adder. *E-Print: quant-ph/0406142*, 2004, quant-ph/0406142. 77, 91, 124

[70] T. Draper, S. Kutin, E. Rains, and K. Svore. A logarithmic-depth quantum carry-lookahead adder. *ArXiv Quantum Physics e-prints*, 2004, arXiv:quant-ph/0406142. 127

[71] L.M. Duan, B.B. Blinov, D.L. Moehring, and C. Monroe. Scalable trapped ion quantum computation with a probabilistic ion-photon mapping. *E-Print: quant-ph/0401020*, 2004, quant-ph/0401020. 53, 80, 160, 162, 164

[72] W. Dur, H.J. Briegel, J.I. Cirac, and P. Zoller. Quantum repeaters based on entanglement purification. *Phys. Rev.*, A59:169, 1999. DOI: 10.1103/PhysRevA.59.169 82, 110, 113

[73] R.J. Elliott, P. Pfeuty, and C. Wood. Ising model with a transverse field. *Phys. Rev. Lett.*, 25(7):7, 1970. DOI: 10.1103/PhysRevLett.25.443 149

[74] E. Farhi, J. Goldstone, S. Gutmann, and M. Sipser. Quantum computation by adiabatic evolution. *ArXiv.org:quant-ph/0001106*, 2000, quant-ph/0001106. 21

[75] E. Farhi, J. Goldstone, S. Gutmann, and M. Sipser. Quantum computation by adiabatic evolution. *ArXiv Quantum Physics e-prints*, 2000, quant-ph/0001106v1. 23, 45, 46, 47

[76] E. Farhi, J. Goldstone, and S. Gutmann. A numerical study of the performance of a quantum adiabatic evolution algorithm for satisfiability. *ArXiv Quantum Physics e-prints*, 2000, quant-ph/0007071v1. 47

[77] A.G. Fowler, W.F. Thompson amd Z. Yan, A.M. Stephens, B.L.T. Plourde, and F.K. Wilhelm. Long-range coupling and scalable architecture for superconducting flux qubits. *arXiv:cond-mat/0702620*, 2007. DOI: 10.1103/PhysRevB.76.174507 83

[78] G.A. Fowler, W.F. Thompson, Z. Yan, A.M. Stephens, B.L.T. Plourde, and Frank K. Wilhelm. Long-range coupling and scalable architecture for superconducting flux qubits. *Phys. Rev. B.*, 76(174507), 2007. DOI: 10.1103/PhysRevB.76.174507 50, 53, 138

[79] M.H. Freedman, A. Kitaev, M.J. Larsen, and Z. Wang. Topological quantum computation. *Bulletin of the American Mathematical Society*, 40:3138, 2003, quant-ph/0101025. 21

[80] A. Furusawa, J. Sorensen, S.L. Braunstein, C. Fuchs, H.J. Kimble, and E.S. Polzik. Unconditional quantum teleportation. *Science*, 282:706709, 1998. DOI: 10.1126/science.282.5389.706 53

[81] G. Gelfond. MagiQ technologies. *www.magiqtech.com*, 2002. 3

[82] N.A. Gershenfeld and I.L. Chuang. Bulk spin-resonance quantum computation. *Science*, 275(350):350–356, 1997. DOI: 10.1126/science.275.5298.350 52

[83] N. Gisin, G. Ribordy, W. Tittel, and H. Zbinden. Quantum cryptography. *Reviews of Modern Physics*, 74:145195, 2002. DOI: 10.1103/RevModPhys.74.145 53

[84] D. Gottesman. A class of quantum error-correcting codes saturating the quantum hamming bound. *Phys. Rev. A*, 54:1862, 1996, quant-ph/9604038. DOI: 10.1103/PhysRevA.54.1862 57, 86, 89

[85] D. Gottesman. Theory of fault-tolerant quantum computation. *Phys. Rev. A*, 57:127–137, 1998, quant-ph/9702029. DOI: 10.1103/PhysRevA.57.127 72, 89

[86] D. Gottesman. Fault tolerant quantum computation with local gates. *Journal of Modern Optics*, 47:333–345, 2000, quant-ph/9903099. DOI: 10.1103/PhysRevA.57.127 57, 72, 74, 76, 131

[87] D.K. Gottesman and I.L. Chuang. Quantum teleportation is a universal computational primitive. *Nature*, (402):390–392, 1999, quant-ph/9908010. DOI: 10.1038/46503 17, 70, 94, 132, 159, 162, 163

[88] D. Greenberger, M. Horne, A. Shimony, and Zeilinger. Bell's theorem without the inequalities. *Amer. J. Phys*, 58:1131–1143, 1990. The GHZ state inventors. DOI: 10.1119/1.16243 162

[89] L. Grover. A fast quantum mechanical algorithm for database search. *Symposium on Theory of Computing (STOC 1996)*, pages 212–219, 1996. DOI: 10.1145/237814.237866 3, 23, 38, 97

[90] E.L. Hahn. Spin echoes. *Phys. Rev.*, 80:580–594, 1950. DOI: 10.1103/PhysRev.80.580 54

[91] S. Hallgren. Polynomial time quantum algorithms or Pell's equation and the principal ideal problem. *Symposium on Theory of Computing (STOC 2002)*, pages 653–658, May 2002. DOI: 10.1145/1206035.1206039 3

[92] A. Harrow, P. Hayden, and D. Leung. Superdense coding of quantum states. *Phys. Rev. Lett.*, 92(187901), 2004, arXiv:quant-ph/0307221. DOI: 10.1103/PhysRevLett.92.187901 19

[93] S. Hellberg. Robust quantum computation with quantum dots. *arXiv.org: quant-ph/0304150*, 2003, quant-ph/0304150. 138

[94] J.L. Hennessy and D.A. Patterson. *Computer Architecture: A Quantitative Approach (The Morgan Kaufmann Series in Computer Architecture and Design)*. Morgan Kaufman Publishers, 340 Pine Street, San Francisco, CA 94104, 2003. 57

[95] W.K. Hensinger, S. Olmschenk, D. Stick, D. Hucul, M. Yeo, M. Acton, L. Deslauriers, J. Rabchuk, and C. Monroe. T-junction ion trap array for two-dimensional ion shuttling, storage and manipulation. *E-Arxiv: quant-ph/0508097*, 2005. DOI: 10.1063/1.2164910 55, 104

[96] T. Hime, P.A. Reichardt, B.L.T. Plourde, T.L. Robertson, C.-E. Wu, A.V. Ustinov, and John Clarke. Solid-state qubits with current-controlled coupling. *Science*, 314(5804):1427 – 1429, 2007. DOI: 10.1126/science.1134388 51, 83

[97] T. Hogg, C. Mochon, W. Polak, and E. Rieffel. Tools for quantum algorithms. *Int. J. Mod. Phys.*, C10:1347–1362, 1999, quant-ph/9811073. 134

[98] L.C.L. Hollenberg, A.S. Dzurak, C. Wellard, A.R. Hamilton, D.J. Reilly, G.J. Milburn, and R.G. Clark. Charge-based quantum computing using single donors in semiconductors. *Phys. Rev. B*, 69(113301), 2003, cond-mat/0306235. DOI: 10.1103/PhysRevB.69.113301 52, 83, 138

[99] J.C. Howell, J.A. Yeazell, and D. Ventura. Optically simulating a quantum associative memory. *Phys. Rev. A*, 62(042303), 2000. DOI: 10.1103/PhysRevA.62.042303 53

[100] Y. Hu and A. Du. Magnetization behavior and magnetic entropy change of frustrated Ising antiferromagnets on two and three dimensional lattices. *J. Phys.:Condens. Matter*, 20:125225, 2008. DOI: 10.1088/0953-8984/20/12/125225 149

[101] N. Isailovic, Y. Patel, M. Whitney, and J. Kubiatowicz. Interconnection networks for scalable quantum computers. *International Symposium of Computer Architecture (ISCA-33), Boston, MA*, 2006. DOI: 10.1109/ISCA.2006.24 113, 169

[102] N. Isailovic, M. Whitney, Y. Patel, J. Kubiatowicz, D. Copsey, F.T. Chong, I.L. Chuang, and M. Oskin. Datapath and control for quantum wires, appears in *Transactions on Architecture and Code Optimization (taco)* 1(1):34–61, 2004. DOI: 10.1145/980152.980155 169

[103] N. Isailovic, M. Whitney, Y. Patel, and J. Kubiatowicz. Running a quantum circuit at the speed of data. In *International Symposium on Computer Architecture (ISCA)*, pages 177–188, 2008. DOI: 10.1109/ISCA.2008.5 2, 126, 127, 128, 169

[104] K. Iwama, Y. Kambayashi, and S. Yamashita. Transformation rules for designing cnot-based quantum circuits. *Proc. Design Automation Conference (DAC)*, pages 419–424, 2002, quant-ph/0401162. DOI: 10.1145/513918.514026 133, 135

[105] J.A. Jones, V. Vedral, A. Ekert, and G. Castagnoli. Geometric quantum computation using nuclear magnetic resonance. *Nature*, 403:869871, 2000. DOI: 10.1038/35003101 21

[106] R. Jullien, P. Pfeuty, J.N. Fields, and S. Doniach. Zero-temperature renormalization method for quantum systems. I. Ising model in a transverse field in one dimension. *Phys. Rev. B*, 18(7):3568–3578, Oct 1978. DOI: 10.1103/PhysRevB.18.3568 149

[107] A. Juozapavicius, S. Caprara, and A. Rosengren. Quantum ising model in a transverse random field: A density-matrix renormalization-group analysis. *Phys. Rev. B*, 56(17):11097–11101, Nov 1997. DOI: 10.1103/PhysRevB.56.11097 149, 150

[108] B.E. Kane. A silicon-based nuclear spin quantum computer. *Nature*, 393:133–137, 1998. DOI: 10.1038/30156 52, 59, 80, 83

[109] B.E. Kane. Silicon based quantum computation. *Progress of Physics*, 48:1023–1041, 2000. 52, 138

[110] P. Kaye, R. Laflamme, and M. Mosca. *An Introduction to Quantum Computing*. Oxford University Press, Oxford, 2007. 23, 24, 25, 37, 44

[111] J. Kempe, A. Kitaev, and O. Regev. The complexity of the local hamiltonian problem. *SIAM Journal of Computing*, 35:1070–1097, 2006. DOI: 10.1137/S0097539704445226 151

[112] D. Kielpinski, C. Monroe, and D.J. Wineland. Architecture for a large-scale ion-trap quantum computer. *Nature*, 417:709–711, 2002. DOI: 10.1038/nature00784 52, 54, 55, 120

[113] Y.-H Kim, S.P. Kulik, and Y. Shih. Quantum teleportation of a polarization state with a complete bell state measurement. *Phys. Rev. Lett.*, 86:13701373, 2001. DOI: 10.1103/PhysRevLett.86.1370 53

[114] J. Kim, S. Pau, Z. Ma, H.R. McLellan, J.V. Gates, A. Kornblit, and R.E. Slusher. System design for a large-scale ion-trap quantum information processor. *Quantum Information and Computation*, 5 (7):515, 2005. 1, 54, 55, 56, 105, 131, 143

[115] A. Kitaev. Quantum measurements and the abelian stabilizer problem. *ArXiv Quantum Physics e-prints*, 1995, arXiv:quant-ph/9511026v1. 24, 26

[116] A.Y. Kitaev. Quantum error correction with imperfect gates. *3rd Int. Conf. of Quantum Communication and Measurement*, pages 181–188, 1997. 57

[117] A. Yu. Kitaev, A.H. Shen, and M.N. Vyalyi. *Classical and Quantum Computation*, volume 47 of *Graduate Studies in Mathematics*. American Mathematical Society, Providence, 2002. 154

[118] E. Knill. Conventions for quantum pseudocode. *Los Alamos National Laboratories, Technical Report LAUR-96-2724*, 1996. 49

[119] E. Knill. Quantum gates using linear optics and postselection. *Phys. Rev. A*, 66, 2002. DOI: 10.1103/PhysRevA.66.052306 53, 167

[120] E. Knill. Quantum computing with very noisy devices. *http://arxiv.org/abs/quant-ph/0410199*, 2004, quant-ph/0410199. 53, 167, 169

[121] E. Knill. Quantum computing with reallistically noisy devices. *Nature*, 434:39–44, 2005. DOI: 10.1038/nature03350 138

[122] E. Knill and R. Laflamme. A theory of quantum error-correcting codes. *Phys. Rev. A*, 55:900–911, 1997, quant-ph/9604034. 57

[123] E. Knill, R. Laflamme, and G.J. Milburn. A scheme for efficient quantum computation with linear optics. *Nature*, 409:4652, 2001. DOI: 10.1038/35051009 52, 53, 167, 169

[124] E. Knill, R. Laflamme, and W. Zurek. Threshold accuracy for quantum computation. *Online Physics Archive: arXiv.org:quant-ph/9610011*, 1996. 159

[125] L. Kreger-Stickles and M. Oskin. Microcoded architectures for ion-tap quantum computers. In *International Symposium on Computer Architecture (ISCA)*, pages 165–176, 2008. DOI: 10.1109/ISCA.2008.28 99

[126] P. Kwiat, K. Mattle, H. Weinfurter, A. Zeilinger, A.V. Sergienko, and Y.H. Shih. New high-intensity source of polarization-entangled photon pairs. *Phys. Rev. Lett.*, 75:43374341, 1995. DOI: 10.1103/PhysRevLett.75.4337 53

[127] P.A. Kwiat, J.R. Mitchell, P.D.D. Schwindt, and A.G. White.   Grovers search algorithm: An optical approach.   *Journal of Modern Optics*, 47:257266, 2000. DOI: 10.1080/09500340008244040 53

[128] P. Kwiat, E. Waks, A.G. White, I. Appelbaum, and P.H. Eberhard.   Ultra-bright source of polarization-entangled photons.   *Phys. Rev. A*, 60:773–776, 1999. DOI: 10.1103/PhysRevA.60.R773 53

[129] T.D. Ladd, J.R. Goldman, F. Yamaguchi, Y. Yamamoto, E. Abe, and K.M. Itoh.   An all silicon quantum computer. *(Physics Online Arxive) arXiv.org:quant-ph/0109039*, 2001. DOI: 10.1103/PhysRevLett.89.017901 52

[130] C. Langer and et.al. Long-lived qubit memory using atomic ions. *E-Print: quant-ph/0504076*, 2005, quant-ph/0504076. DOI: 10.1103/PhysRevLett.95.060502 105, 141

[131] J. Lantz, M. Wallquist, V.S. Shumeiko, and G. Wendin. Josephson junction qubit network with current-controlled interaction. *Phys. Rev. B*, 70(140507(R)):66–70, 2004, cond-mat/0403285. DOI: 10.1103/PhysRevB.70.140507 80

[132] D. Leibfried and et al. Experimental demonstration of a robust, high-fidelity geometric two ion-qubit phase gate. *Nature*, 422:412–415, 2003. DOI: 10.1038/nature01492 54

[133] D.A. Lidar and L.-A Wu. Encoded recoupling and decoupling: An alternative to quantum error correcting codes, applied to trapped ion quantum computation. *Phys. Rev. A.*, 67:032313, 2003. DOI: 10.1103/PhysRevA.67.032313 60

[134] H.K. Lo and H.F. Chau.   Unconditional security of quantum key distribution.   *Science*, 283:2050–2056, 1999, quant-ph/9803006. DOI: 10.1126/science.283.5410.2050 3

[135] C.-Y. Lu, D.E. Browne, T. Yang, and J.-P. Pan.   Demonstration of shor's quantum factoring algorithm using photonic qubits.   *arXiv:0705.1684v1 [quant-ph]*, 2007. DOI: 10.1103/PhysRevLett.99.250504 51

[136] Y. Makhlin, G. Schoen, and A. Shnirman. Josephson-junction qubits with controlled couplings. *Nature*, 398:305, 1999. DOI: 10.1038/18613 52, 80

[137] D. Maslov, S.M. Falconer, and M. Mosca. Quantum circuit placement: Optimizing qubit-to-qubit interactions through mapping quantum circuits into a physical experiment. *E-Print arXiv:quant-ph/0703256v1*, 2007. DOI: 10.1145/1278480.1278717 133

[138] D.N. Matsukevich and A. Kuzmich. Quantum state transfer between matter and light. *Science*, 306(5696):663666, 2004. DOI: 10.1126/science.1103346 53, 80, 160, 164

[139] L. McMurchie and C. Eberling. Pathfinder: A negotiation based performance-driven router for fpgas. *in Proceedings of ACM Symp. on Field Programmable Gate Arrays*, pages 111–117, 1995. DOI: 10.1145/201310.201328 131

[140] T.S. Metodi, Darshan D. Thaker, Andrew W. Cross, Frederic T. Chong, and Isaac L. Chuang. A quantum logic array microarchitecture: Scalable quantum data movement and computation. *Proceedings of the 38th International Symposium on Microarchitecture*, MICRO-38, 2005. DOI: 10.1109/MICRO.2005.9 xii, 1, 2, 50, 82, 97, 116, 117, 120

[141] T.S. Metodi, Darshan D. Thaker, Andrew W. Cross, Frederic T. Chong, and Isaac L. Chuang. Physical operations scheduler in a quantum information processor. *Proceedings of the SPIE Defense and Security Symposium, Orlando FL*, 2006. DOI: 10.1117/12.666419 131, 138, 140, 160

[142] K. Michielsen and Hans De Raedt. QCE: A simulator for quantum computer hardware. *Turk J Phys*, 27:343, 2003. DOI: 10.1088/0957-4484/13/1/305 85

[143] G.J. Milburn. Quantum optical fredkin gate. *Phys. Rev. Lett.*, 62:2124–2127, 1989. DOI: 10.1103/PhysRevLett.62.2124 52

[144] R. Moessner and S.L. Sondhi. Ising and dimer models in two and three dimensions. *Physical Review B.*, 68:054405, 2003. DOI: 10.1103/PhysRevB.68.054405 149

[145] C. Monroe. Quantum information processing with atoms and photons. *Nature*, 416:238, 2002. DOI: 10.1038/416238a 20, 53, 160, 162, 164

[146] C. Monroe, C.A. Sackett, D. Kielpinski, B.E. King, C. Langer, V. Meyer, C.J. Myatt, M. Rowe, Q.A. Turchette, W.M. Itano, and D.J. Wineland. Scalable entanglement of trapped ions. *Workshop on Trapped Ion Quantum Computing (NIST, Boulder, Colorado)*, 2000. 108

[147] T. Nakassis, J.C. Beinfang, P. Johnson, A. Mink, D. Rogers, X. Tang, and C.J. Williams. Has quantum cryptography been proven secure. *Proceedings of the SPIE Defense and Security Symposium, Orlando FL*, 2006. DOI: 10.1117/12.665086 3

[148] M. Nielsen. Optical quantum computation using cluster states. *Phys. Rev. Lett*, 93 (040503), 2004, quant-ph/0402005. DOI: 10.1103/PhysRevLett.93.040503 21, 53, 54

[149] M.A. Nielsen and I.L. Chuang. *Quantum Computation and Quantum Information*. Cambridge University Press, Cambridge, UK, 2000. 18, 23, 30, 35, 53, 57, 74, 85, 98, 135, 152

[150] M.A. Nielsen and C.M. Dawson. Fault-tolerant quantum computation with cluster states. *arXiv.org:quant-ph/0405134*, 2004, arXiv.org:quant-ph/0405134. DOI: 10.1103/PhysRevA.71.042323 21, 54

[151] A.O. Niskanen, K. Harrabi, F. Yoshihara, Y. Nakamura, S. Lloyd, and J.S. Tsai. Quantum coherent tunable coupling of superconducting qubits. *Science*, 316(5825):723 – 726, 2007. DOI: 10.1126/science.1141324 51, 83

[152] B. Omer. A procedural formalism for quantum computing: Qcl. *Master thesis technical physics, TU Vienna*, 1998. 85, 132

[153] M. Oskin, F. Chong, and I. Chuang. A practical architecture for reliable quantum computers. *IEEE Computer*, 35:79–87, 2002. DOI: 10.1109/2.976922 1, 50, 86, 169

[154] M. Oskin, F.T. Chong, J. Kubiatowicz, and I.L. Chuang. Building quantum wires: The long and the short of it. *in Proc. ISCA-30, San Diego, CA*, 2003. DOI: 10.1109/ISCA.2003.1207015 50, 79, 138, 169

[155] Z.Y. Ou and L. Mandel. Violation of bell's inequality and classical probability in a twophoton correlation experiment. *Phys. Rev. Lett.*, 61:50–53, 1988. DOI: 10.1103/PhysRevLett.61.50 53

[156] R. Ozeri and et. al. Hyperfine coherence in the presence of spontaneous photon scattering. *arXiv:quant-ph/0502063*, 2004. DOI: 10.1103/PhysRevLett.95.030403 54, 66, 104, 105

[157] J.-W. Pan, D. Bouwmeester, H. Weinfurter, and A. Zeilinger. Experimental entanglement swapping: Entangling photons that never interacted. *Phys. Rev. Lett.*, 80:38913894, 1998. DOI: 10.1103/PhysRevLett.80.3891 53

[158] S. Parker and M.B. Plenio. Efficient factorization with a single pure qubit and logn mixed qubits. *Phys. Rev. Lett.*, 85(3049):3049, 2000. DOI: 10.1103/PhysRevLett.85.3049 151

[159] K.N. Patel, I.L. Markov, and J.P. Hayes. Efficient synthesis on linear reversible circuits. *ArXiv Quantum Physics e-prints*, 2003, quant-ph/0302002. 133, 170

[160] P. Pfeuty. The one-dimensional Ising model with a transverse field. *Annals of Physics*, 57:79–90, March 1970. DOI: 10.1016/0003-4916(70)90270-8 149, 150, 151

[161] P.M. Platzman and M.I. dykman. Quantum computing with electrons floating on liquid helium. *Science*, 284:19671969, 1999. DOI: 10.1126/science.284.5422.1967 52

[162] A. Politi, J.C.F. Matthews, and J.L. O'Brien. Shor's quantum factoring algorithm on a photonic chip. *Science*, 325(1221), 2009. DOI: 10.1126/science.1173731 53

[163] J.V. Porto, S. Rolston, T.B. Laburthe, C.J. Williams, and W.D. Phillips. Quantum information with neutral atoms as qubits. *Phil. Trans. R. Soc. Lond.*, A361:1417–1427, 2003. DOI: 10.1098/rsta.2003.1211 54

[164] D. Poulin. Stabilizer formalism for operator quantum error correction. 2005, quant-ph/0508131. DOI: 10.1103/PhysRevLett.95.230504 94, 131, 169

[165] V. Privman, I.D. Vagner, and G. Kventsel. Quantum computation in quantum-hall systems. *Physics Letters A*, 239:141, 1998. DOI: 10.1016/S0375-9601(97)00974-2 52

[166] P. Rabl, D. DeMille, J.M. Doyle, M.D. Lukin, R.J. Schoelkopf, and P. Zoller. Hybrid quantum processors: Molecular ensembles as quantum memory for solid state circuits. *Phys. Rev. Lett.*, 97(033303), 2006. DOI: 10.1103/PhysRevLett.97.033003 51

[167] M. Reck, A. Zeilinger, H.J. Bernstein, and P. Bertani. Experimental realization of any discrete unitary operator. *Phys. Rev. Lett.*, 73:58–61, 1998. DOI: 10.1103/PhysRevLett.73.58 53

[168] B.W. Reichardt. Improved ancilla preparation scheme increases fault-tolerant threshold. *E-Print: quant-ph/0406025*, 2004, quant-ph/0406025. 75, 106, 107

[169] R. Reichle1, D. Leibfried, E. Knill, J. Britton, R.B. Blakestad, J.D. Jost, C. Langer, R. Ozeri, S. Seidelin, and D.J. Wineland. Experimental purification of two-atom entanglement. *Nature*, 443(19):838–841, 2006. DOI: 10.1038/nature05146 51, 141

[170] G. Ribordy, O. Guinnard, and H. Zbinden. id Quantique. *www.idquantique.com*, 2004. 3

[171] M. Riebe, H. Haffner, C.F. Roos, and et. al. Deterministic quantum teleportation with atoms. *Nature*, 429(6993):734–737, 2004. DOI: 10.1038/nature02570 51, 52, 54, 79, 108, 141, 160

[172] R. Rivest, A. Shamir, and L. Adleman. A method for obtaining digital signatures and public-key cryptosystems. *Communications of the ACM*, 21(2):120126, 1978. DOI: 10.1145/359340.359342 3, 23

[173] D. Rosn, J. Olsson, and C. Hedlund. Membrane covered electrically isolated through-wafer via holes. *J Micromech. Microeng*, 11:344, 2001. DOI: 10.1088/0960-1317/11/4/310 56

[174] M.A. Rowe and et. al. Transport of quantum states and separation of ions in a dual rf ion trap. *Quant. Inf. Comp.*, 2:257–271, 2002. 55

[175] S. Sachdev. *Quantum Phase Transitions*. Cambridge University Press, Cambridge, England, U.K., 1999. 149, 150

[176] M. Santos. Short-time critical dynamics for the transverse ising model. *Phys. Rev. E*, 61(6):7204–7207, Jun 2000. DOI: 10.1103/PhysRevE.61.7204 150

[177] B. Schumacher. Quantum coding. *Phys. Rev. A*, 51:27382747, 1995. DOI: 10.1103/PhysRevA.51.2738 2

[178] S. Seidelin, J. Chiaverini, R. Reichle, J.J. Bollinger, and et.al. A microfabricated surface-electrode ion trap for scalable quantum information processing. *ArXiv Quantum Physics e-prints*, 2006, quant-ph/0601173. DOI: 10.1103/PhysRevLett.96.253003 55, 140, 141

[179] V.V. Shende, S.S. Bullock, and I.L. Markov. Synthesis of quantum logic circuits. *IEEE Trans. on Computer-Aided Design*, 25(6):1000–1010, 2006. DOI: 10.1109/TCAD.2005.855930 83

[180] V.V. Shende, S.S. Bullock, and I.L. Markov. Recognizing small-circuit structure in two-qubit operators and timing hamiltonians to compute controlled-not gates. *Phys. Rev. A*, (012310), 2003, quant-ph/0308045. DOI: 10.1103/PhysRevA.70.012310 134

[181] V.V. Shende and I.L. Markov. Quantum circuits for incompletely specified two-qubit operators. *Quantum Information and Computation*, (5):048–056, 2005. 170

[182] V.V. Shende, I.L. Markov, and S.S. Bullock. Minimal universal two-qubit quantum circuits. *Phys. Rev. A*, 69(062321):1–7, 2003, quant-ph/0308033. 134

[183] V.V. Shende, I.L. Markov, and S.S. Bullock. Finding small two-qubit circuits. *Proceedings of the SPIE*, (5436):348–359, 2004. DOI: 10.1117/12.542381 134, 170

[184] P.W. Shor. Polynomial-time algorithms for prime factorization and discrete logarithms on a quantum computer. *35th Annual Symposium on Foundations of Computer Science*, pages 124–134, 1994. DOI: 10.1109/SFCS.1994.365700 3, 19, 23, 30, 76, 91, 131

[185] P.W. Shor. Scheme for reducing decoherence in quantum computer memory. *Phys. Rev. A*, 54:2493, 1995. DOI: 10.1103/PhysRevA.52.R2493 57, 62, 66, 94

[186] P.W. Shor. Fault-tolerant quantum computation. *in Proc. 37th Symp. on Foundations of Computer Science*, 1996. 159

[187] D. Simon. On the power of quantum computation. *Proceedings of the 35th Annual Symposium on Foundations of Computer Science IEEE Computer Society Press, Los Alamitos, CA*, pages 116–123, 1994. DOI: 10.1109/SFCS.1994.365701 19

[188] A.J. Skinner, M.E. Davenport, and B.E. Kane. Hydrogenic spin quantum computing in silicon: A digital approach. *Phys. Rev. L*, 90(087901), February 2003, quant-ph/0206159. DOI: 10.1103/PhysRevLett.90.087901 52, 83

[189] G. Song and A. Klappenecker. Optimal realizations of controlled unitary gates. *ournal of Quantum Information and Computation*, 3(2):139–155, 2003, quant-ph/0207157. 17

[190] G. Song and A. Klappenecker. Optimal realizations of controlled unitary gates. *ournal of Quantum Information and Computation*, 3(2):139–155, 2003, quant-ph/0207157. 134

[191] A. Sorensen and K. Molmer. Entanglement and quantum computation with ions in thermal motion. *Phys. Lett. A*, 62:02231, 2000. DOI: 10.1103/PhysRevA.62.022311 54, 108

[192] T.P. Spiller, K. Nemoto, S.L. Braunstein, W.J. Munro, P. van Loock, and G.J. Milburn. Quantum computation by communication. *http://arxiv.org/abs/quant-ph/0509202*, 2005, quant-ph/050902. DOI: 10.1088/1367-2630/8/2/030 53, 80, 160, 164

[193] A.M. Steane. How to build a 300 bit, 1 gop quantum computer. *arXiv:quant-ph/0412165*, 2004, quant-ph/0412165. 54, 55, 56, 57, 105, 141

[194] A. Steane. Error correcting codes in quantum theory. *Phys. Rev. Lett*, 77:793–797, 1996. DOI: 10.1103/PhysRevLett.77.793 57, 67

[195] A.M. Steane. Overhead and noise threshold of fault-tolerant quantum error correction. *E-Print: quant-ph/0207119*, 2002, quant-ph/0207119. DOI: 10.1103/PhysRevA.68.042322 57, 60, 75, 76

[196] A.M. Steane. Space, time, parallelism and noise requirements for reliable quantum computing. *Fortsch. Phys.*, 46:443–458, 1998, quant-ph/9708021. DOI: 10.1002/(SICI)1521-3978(199806)46:4/5%3C443::AID-PROP443%3E3.0.CO;2-8 66, 68

[197] A.M. Steane. Efficient fault-tolerant quantum computing. *Phys. Rev. Lett.*, 78:2252–2255, 1997, quant-ph/9809054. DOI: 10.1038/20127 159

[198] M. Suzuki. General theory of high order decomposition of exponential operators and symplectic intergration. *Phys. Lett. A*, 165:387, 1992. DOI: 10.1016/0375-9601(92)90335-J 152

[199] K.M. Svore, A.W. Cross, A.V. Aho, I.L. Chuang, and I.L. Markov. Toward a software architecture for quantum computing design tools. *Workshop on Quantum Programming Languages (QPL)*, 2004. 74, 170

[200] K.M. Svore, A.W. Cross, I.L. Chuang, and A. Aho. Pseudothreshold or threshold? - more realistic threshold estimates for fault-tolerant quantum computing. *E-Print: quant-ph/0508176*, 2005, quant-ph/0508176. 77, 80, 170

[201] K.M. Svore, D.P. DiVincenzo, and B.M. Terhal. Noise threshold for a fault-tolerant two-dimensional lattice architecture. *E-Print (Arxiv.org): quant-ph/0604090*, 2006, quant-ph/0604090. 69, 70, 74, 80, 83, 131, 138, 139

[202] K.M. Svore, B. Terhal, and D.P. DiVincenzo. Local fault-tolerant quantum computation. *E-Print: quant-ph/0410047*, 2004, quant-ph/0410047. DOI: 10.1103/PhysRevA.72.022317 107, 131

[203] S. Takeuchi. Analysis of errors in linear-optics quantum computation. *Phys. Rev. A*, 61(052302), 2000. DOI: 10.1103/PhysRevA.61.052302 53

[204] J.M. Taylor, H.A. Engel, W. Dur, A. Yacoby, C.M. Marcus, P. Zoller, and M.D. Lukin. Fault-tolerant architecture for quantum computation using electrically controlled semiconductor spins. *Nature Physics*, 1:177–183, 2005. DOI: 10.1038/nphys174 51

[205] T. Toffoli. *Reversible Computing*. Springer, New York, 2000. 15

[206] Q.A. Turchette, C.J. Hood, W. Lange, H. Mabuchi, and H.J. Kimble. Measurement of conditional phase shifts for quantum logic. *Physical Review Letters*, 75:4710, 1995. DOI: 10.1103/PhysRevLett.75.4710 52

[207] M. Udrescu, L. Prodan, and M. Vladutiu. Using hdls for describing quantum circuits: A framework for efficient quantum algorithm simulation. *2nd ACM International Conference on Computing Frontiers (CF'05), Ischia, Italy*, pages 96–110, 2004. DOI: 10.1145/977091.977107 169

[208] M. Udrescu, L. Prodan, and M. Vladutiu. Improving quantum circuit dependability with reconfigurable quantum gate arrays. *2nd ACM International Conference on Computing Frontiers (CF'05), Ischia, Italy*, pages 133–144, 2005. DOI: 10.1145/1062261.1062286 169

[209] L.G. Valiant. Quantum circuits that can be simulated classically in polynomial time. *Proc. ACM Symposium on Theory of Computing (STOC)*, page 114, 2001. DOI: 10.1137/S0097539700377025 85

[210] W. van Dam, M. Mosca, and U.V. Vazirani. How powerful is adiabatic quantum computation? In *Symposium on Foundations of Computer Science (FOCS)*, pages 279–287, 2001. DOI: 10.1109/SFCS.2001.959902 47

[211] W. van Dam and G. Seroussi. Efficient quantum algorithms for estimating gauss sums. *E-Print: quant-ph/0207131*, 2002, quant-ph/0207131. 3

[212] L.M.K. Vandersypen, M. Steffen, G. Breyta, C.S. Yannoni, M. Sherwood, and I.L. Chuang. Experimental realization of shor's quantum factoring algorithm using nuclear magnetic resonance. *Nature*, 414:883, 2001. DOI: 10.1038/414883a 1

[213] R. van Meter and et.al. Distributed arithmetic on a quantum multi-computer. *International Symposium of Computer Architecture (ISCA-33), Boston, MA*, 2006. DOI: 10.1109/ISCA.2006.19 138, 163, 164, 165, 169

[214] R. van Meter and K.M. Itoh. Fast quantum modular exponentiation. *E-Print: quant-ph/0408006*, 2004, quant-ph/0408006. DOI: 10.1103/PhysRevA.71.052320 77

[215] R. van Meter and M. Oskin. Architectural implications of quantum computing technologies. *ACM Journal on Emerging Technologies in Computing Systems (JETC)*, 2(1):31–63, 2006. DOI: 10.1145/1126257.1126259 52, 80, 104

[216] R. van Meter and M. Oskin. Architectural implications of quantum computing technologies. *ACM Journal on Emerging Technologies in Computing Systems (JETC)*, 2(1):31–68, 2006. DOI: 10.1145/1126257.1126259 169

[217] G.F. Viamontes, I.L. Markov, and J.P. Hayes. Graph-based simulation of quantum computation in the density matrix representation. *Quantum Information and Computation*, 5(2):113–130, 2005, quant-ph/0403114. DOI: 10.1117/12.542767 85

[218] J. von Neuman. Probabilistic logic and the synthesis of reliable organisms from unreliable components. *Automata Series, Editors: C. Shannon and J. McCarthy, Princeton Univ. Press*, pages 43–98, 1956. 65

[219] G. Vidal. Efficient classical simulation of slightly entangled quantum computations. *Physical Review Letters*, 91:147902, 2003. DOI: 10.1103/PhysRevLett.91.147902 85

[220] M. Whitney, N. Isailovic, Y. Patel, and J. Kubiatowicz. A fault tolerant, area efficient architecture for shor's factoring algorithm. In *Proceedings of International Symposium on Computer Architecture (ISCA)*, pages 383–394, 2009. DOI: 10.1145/1555754.1555802 2, 101, 169

[221] D.J. Wineland and et al. Experimental issues in coherent quantum-state manipulation of trapped atomic ions. *Journal of Research of NIST*, 103:259–328, 1998, quant-ph/9710025. 52, 54

[222] D. Wineland and T. Heinrichs. Ion trap approaches to quantum information processing and quantum computing. *A Quantum Information Science and Technology Roadmap*, 2004. URL: http://quist.lanl.gov. 53, 54, 104

[223] D.J. Wineland, D. Leibfried, M.D. Barrett, A. Ben-Kish, and et.al. Quantum control, quantum information processing, and quantum-limited metrology with trapped ions. *Proceedings of the International Conference on Laser Spectroscopy (ICOLS)*, 2005, quant-ph/0508025. DOI: 10.1142/9789812701473_0040 104

[224] F.A. Wolf. *Taking the Quantum Leap*. Harper and Snow, San Francisco, CA, 1981. 20

[225] W.K. Wootters and W.H. Zurek. A single quantum cannot be cloned. *Nature*, 299:802–803, 1982. DOI: 10.1038/299802a0 14

[226] A. Yao. Quantum circuit complexity. *Proceedings of the 34th Annual Symposium on Foundations of Computer Science*, pages 352–361, 1993. DOI: 10.1109/SFCS.1993.366852 80

[227] B. Zeng, D.L. Zhou, Z. Xu, and C.P. Sun. Quantum teleportation using cluster states. *ArXiv Quantum Physics e-prints (arXiv:quant-ph/0304165)*, 2003, arXiv:quant-ph/0304165. 21, 53, 54

[228] B. Zeng, A. Cross, and I.L. Chuang. Transversality versus universality for additive quantum codes. *arXiv:0706.1382v1 [quant-ph]*, 2007. 70

[229] X. Zhou, D. Leung, and I.L. Chuang. Methodology for quantum logic gate construction. *arXive: quant-ph/0002039*, quant-ph/0002039. DOI: 10.1103/PhysRevA.62.052316 162, 166

Authors' Biographies

TZVETAN S. METODI

Tzvetan S. Metodi is a senior member of the technical staff at the Computer Systems Research Department at the Aerospace Corporation. Tzvetan received his Bachelors degree in physics from the University of California at Davis and PhD in Computer Science also from UC Davis. Tzvetan's current effort in quantum computing focuses on the development of balanced architectural models of organization and specialization for emerging quantum computing technologies, incorporating quantum fault-tolerance and analysis using modern compilation techniques. His other research interests include the design of hardware-based secure partitioning techniques for general-purpose processors running multi-level security flight software systems employed on modern spacecraft.

ARVIN FARUQUE

Arvin Faruque is a graduate student in the Electrical and Computer Engineering department at UC Santa Barbara. He received BS degrees in Computer Engineering and Mathematics from Cal Poly, San Luis Obispo in 2007. His research interests include quantum computation, computer graphics, and applied mathematics.

FREDERIC T. CHONG

Frederic T. Chong is the director of Computer Engineering, the director of the Greenscale Center for Energy-Efficient Computing, and a professor of computer science at the University of California at Santa Barbara. Prof. Chong also leads the computer architecture and circuits areas in both the ORAQL and NGQCS projects under the IARPA Quantum Computer Science program. Prof. Chong also co-founded the Quantum Architecture Research Center (QARC) in 2001, which received the 2002 DARPATech most significant technical achievement award. Prof. Chong received his BS in 1990, MS in 1992, and PhD in 1996, all from MIT. He was an assistant professor at UC Davis from 1996–2001, was an associate professor at UC Davis from 2001–2005, and has been a professor at UCSB from 2005-present. Dr. Chong's research interests include quantum computing architectures, nanoscale electronics, embedded processing, computer security, and sustainable computing. Prof. Chong was a UC Davis Chancellor's Fellow (2002–2007) and received an NSF CAREER Award (1998–2002).

Printed in the United States
by Baker & Taylor Publisher Services